T0336796

Theory of Approximate Functional Equations

Theory of Approximate Functional Equations

in Banach algebras, inner product spaces and amenable groups

Madjid Eshaghi Gordji
Institute for Cognitive Science Studies,
Shahid Beheshti University, Tehran, Iran

Sadegh Abbaszadeh
Intelligent systems and perception recognition
laboratory, CS group of Mathematics
department, Shahid Beheshti University,
Tehran, Iran

AMSTERDAM • BOSTON • HEIDELBERG • LONDON
NEW YORK • OXFORD • PARIS • SAN DIEGO
SAN FRANCISCO • SINGAPORE • SYDNEY • TOKYO
Academic Press is an imprint of Elsevier

Academic Press is an imprint of Elsevier
125 London Wall, London, EC2Y 5AS, UK
525 B Street, Suite 1800, San Diego, CA 92101-4495, USA
50 Hampshire Street, 5th Floor, Cambridge, MA 02139, USA
The Boulevard, Langford Lane, Kidlington, Oxford OX5 1GB, UK

ISBN: 978-0-12-803920-5

Library of Congress Cataloging-in-Publication Data
A catalog record for this book is available from the Library of Congress

British Library Cataloguing in Publication Data
A catalogue record for this book is available from the British Library

For information on all Academic Press publications
visit our website at http://elsevier.com/

Contents

1 **Introduction** **1**

2 **Approximate Cauchy functional equations and completeness** **5**
2.1 Theorem of Hyers 5
2.2 Theorem of Themistocles M. Rassias 9
2.3 Completeness of normed spaces 14

3 **Stability of mixed type functional equations** **21**
3.1 Binary mixtures of functional equations 21
3.2 Ternary mixtures of functional equations 29
3.3 Mixed foursome of functional equations 50

4 **Stability of functional equations in Banach algebras** **69**
4.1 Approximate homomorphisms and derivations in ordinary Banach algebras 69
4.2 Approximate homomorphisms and derivations in C^*-algebras 78
4.3 Stability problem on C^*-ternary algebras 86
4.4 General solutions of some functional equations 99
4.5 Some open problems 107

5 **Stability of functional equations in inner product spaces** **111**
5.1 Introduction 111
5.2 Orthogonal derivations in orthogonality Banach algebras 112
5.3 Some open problems 124

6 **Amenability of groups (semigroups) and the stability of functional equations** **127**
6.1 Introduction 127
6.2 The stability of homomorphisms and amenability 127
6.3 Some open problems 133

Bibliography **137**
Index **141**

Introduction

1

In 1940, an interesting talk presented by Stanislaw M. Ulam triggered the study of stability problems for various functional equations. In his talk, Ulam discussed the notion of the stability of mathematical theorems considered from a rather general point of view: When is it true that by changing a little the hypotheses of a theorem, one can still assert that the thesis of the theorem remains true or approximately true? Particularly, for every general functional equation one can ask the following question: when is it true that the solution of an equation differing slightly from a given one must of necessity be close to the solution of the given equation? Indeed, Ulam proposed the following problem:

Given a group G_1, a metric group G_2 with the metric $d(\cdot,\cdot)$ and a positive number ε, does there exist $\delta > 0$ such that, if a mapping $f : G_1 \to G_2$ satisfies

$$d(f(xy), f(x)f(y)) \leq \delta$$

for all $x, y \in G_1$, then a homomorphism $h : G_1 \to G_2$ exists with

$$d(f(x), h(x)) \leq \varepsilon$$

for all $x \in G_1$? If the problem accepts a solution, the equation is said to be stable (see [90]).

In trying to solve problems of this kind, most authors have considered homomorphisms between different groups or vector spaces, rather than automorphisms, for perturbations.

In the following year, 1941, Donald H. Hyers [6] was able to give a partial solution to Ulam's question that was the first significant breakthrough and step toward more solutions in this area. He considered the case of approximately additive mappings between Banach spaces and proved the following results.

Theorem 1.0.1. *Let E_1 and E_2 be Banach spaces and let $f : E_1 \longrightarrow E_2$ satisfies the following condition: there is a constant $\epsilon \geq 0$ such that*

$$\|f(x+y) - f(x) - f(y)\| \leq \epsilon$$

for all $x, y \in E_1$. Then the limit $h(x) = \lim_{n\to\infty} \frac{f(2^n x)}{2^n}$ exists for all $x \in E_1$ and $h : E_1 \to E_2$ is a unique linear transformation such that

$$\|f(x) - h(x)\| \leq \epsilon.$$

Theory of Approximate Functional Equations. http://dx.doi.org/10.1016/B978-0-12-803920-5.00001-8

Theorem 1.0.2. *If under the hypotheses of Theorem 1.0.1 $f(tx)$ is continuous in $t \in \mathbb{R}$ for each $x \in E_1$, then the mapping h is $\mathbb{R}$-linear.*

The term Hyers-Ulam stability originates from this historical background. The method that was provided by Hyers is called the direct method, and produces the additive mapping h. The direct method is the most important and powerful tool for studying the stability of various functional equations.

The theorem of Hyers was generalized by Aoki [1] for additive mappings by considering an unbounded Cauchy difference.

Theorem 1.0.3 (Aoki [1]). *If $f(x)$ is an approximately linear transformation from E into E', ie,*

$$\|f(x+y)-f(x)-f(y)\| \leq K_0(\|x\|^p + \|y\|^p), \quad \text{for all } x, y \in E_1,$$

where $K_0 \geq 0$ and $0 \leq p < 1$, then there is a unique linear transformation $Q(x)$ near $f(x)$, ie, there exists $K \geq 0$ such that

$$\|f(x) - Q(x)\| \leq K\|x\|^p.$$

In 1978, Themistocles M. Rassias [8] succeeded in extending the result of Hyers's theorem by weakening the condition for the Cauchy difference.

Theorem 1.0.4. *Let E_1 and E_2 be two Banach spaces and let $f : E_1 \longrightarrow E_2$ be a mapping such that $f(tx)$ is continuous in $t \in \mathbb{R}$ for each fixed x. Assume that there exists $\varepsilon > 0$ and $0 \leq p < 1$ such that*

$$\|f(x+y)-f(x)-f(y)\| \leq \varepsilon(\|x\|^p + \|y\|^p), \quad \text{for all } x, y \in E_1.$$

Then there exists a unique linear mapping $T : E_1 \rightarrow E_2$ such that

$$\|f(x) - T(x)\| \leq \frac{2\varepsilon}{2-2^p}\|x\|^p, \quad \text{for all } x \in E_1.$$

The stability phenomenon that was presented by Th.M. Rassias is called the generalized Hyers-Ulam stability. This terminology may also be applied to the cases of other functional equations. Since then, a large number of papers have been published in connection with various generalizations of Ulam's problem and Hyers's theorem.

There are cases in which each approximate mapping is actually a true mapping. In such cases, we say that the functional equation is hyperstable. Indeed, a functional equation is hyperstable if every solution satisfying the equation approximately is an exact solution of it. For the history and various aspects of this theory, we refer the reader to [2–5].

The concept of Hyers-Ulam stability is quite significant in realistic problems in numerical analysis, biology, and economics. This concept actually means that if one is studying a Hyers-Ulam stable system, then one does not have to reach the exact solution (which usually is quite difficult or time consuming). Hyers-Ulam stability guarantees that there is a close exact solution. This is quite useful in many applications, eg, numerical analysis, optimization, biology, and economics, etc., where finding the exact solution is quite difficult. It also helps, if the stochastic effects are small, to use a deterministic model to approximate a stochastic one.

The notion of stability arose naturally in problems of mechanics. There it involves, mathematically speaking, the continuity of the solution of a problem in its dependence on initials parameters. This continuity may be defined in various ways. Often it is sufficient to prove the boundedness of the solutions for arbitrarily long times, eg, the boundedness of the distance between the point representing the system at any time for the initial point, etc. Needless to say, problems of stability occur in other branches of physics and, in a way, also even in pure mathematics.

Unfortunately, there are no books completely dealing with the stability problem of functional equations in Banach algebras, inner product spaces, and amenable groups. Moreover, in most stability theorems for functional equations, the completeness of the target space of the unknown functions contained in the equation is assumed. Recently, the question of whether the stability of a functional equation implies this completeness has been investigated by several authors. In this book, we are going to deal with the above-mentioned developments in the theory of approximate functional equations.

This book introduces the latest and new results on the following topics:

(1) approximate Cauchy functional equations and completeness;
(2) stability of binary, ternary, and foursome mixtures of functional equations;
(3) approximations of homomorphisms and derivations in different Banach algebras;
(4) stability of functional equations in inner product spaces; and
(5) amenability of groups (semigroups) and the stability of functional equations.

Approximate Cauchy functional equations and completeness

2

2.1 Theorem of Hyers

The first result on the stability of functional equations was given in 1941 by Hyers [6]. He was able to give a partial solution to Ulam's question that was the first significant breakthrough and step toward more solutions in this area.

The solutions of the Cauchy (additive) functional equation $f(x+y) = f(x) + f(y)$ have been investigated for many spaces. Any solution $f(x)$ of this equation will be called a linear transformation. To make the statement of the problem precise, let E and E' be Banach spaces and let δ be a positive number. A linear transformation $f(x)$ of E into E' will be called a δ-linear transformation if

$$\|f(x+y) - f(x) - f(y)\| \leq \delta$$

for all x and y in E. Then the problem may be stated as follows. Does there exist for each $\varepsilon > 0$ a $\delta > 0$ such that, to each δ-linear transformation $f(x)$ of E into E' there corresponds a linear transformation $l(x)$ of E into E' satisfying the inequality

$$\|f(x) - l(x)\| \leq \varepsilon$$

for all x in E? Hyers answered this question in the affirmative, and showed that δ may be taken equal to ε. This is clearly a best result, since the transformation $f(x) = L(x)+c$ where $L(x)$ is linear and $c < \varepsilon$ is evidently an ε-linear transformation for which $l(x) = L(x)$.

Theorem 2.1.1 (Hyers [6]). *Let E and E' be Banach spaces and let $f : E \to E'$ be a δ-linear transformation. Then the limit $l(x) = \lim_{n\to\infty} \dfrac{f(2^n x)}{2^n}$ exists for all $x \in E$ and $l : E \to E'$ is the only linear transformation such that*

$$\|f(x) - l(x)\| \leq \varepsilon$$

for all $x \in E$

Theory of Approximate Functional Equations. http://dx.doi.org/10.1016/B978-0-12-803920-5.00002-X

Proof. For any x in E the inequality $\|f(2x) - 2f(x)\| < \delta$ is obvious from the definition of δ-linear transformations. On replacing x by $x/2$ in this inequality and dividing by 2 we see that $\|(1/2)f(x) - f(x/2)\| < \delta/2$ for all x in E. Make the induction assumption

$$\|2^{-n}f(x) - f(2^{-n}x)\| < \delta(1 - 2^{-n}). \tag{2.1}$$

On the basis of the last two inequalities, we find that

$$\|2^{-1}f(2^{-n}x) - f(2^{-n-1}x)\| < \delta/2$$

and

$$\|2^{-n-1}f(x) - 2^{-1}f(2^{-n}x)\| < \delta(1/2 - 2^{-n-1}).$$

Hence,

$$\|2^{-n-1}f(x) - f(2^{-n-1}x)\| < \delta(1 - 2^{-n-1}).$$

Therefore, since the induction assumption [2.1] is known to be true for $n = 1$, it is true for all positive integers n and all $x \in E$.

Put $q_n(x) = f(2^n x)/2^n$, where n is a positive integer and x is in E. Then

$$q_m(x) - q_n(x) = \frac{f(2^m x)}{2^m} - \frac{f(2^n x)}{2^n} = \frac{f(2^{m-n} \cdot 2^n x) - 2^{m-n} f(2^n x)}{2^m}.$$

Therefore, if $m < n$ we can apply inequality [2.1] and write

$$\|q_m(x) - q_n(x)\| < \delta(1 - 2^{m-n})/2^m$$

for all $x \in E$. Hence, $\{q_n(x)\}$ is a Cauchy sequence for each $x \in E$, and since E is complete, there exists a limit function $l(x) = \lim_{n\to\infty} q_n(x)$. Let x and y be any two points of E. Since $f(x)$ is δ-linear,

$$\|f(2^n x + 2^n y) - f(2^n x) - f(2^n y)\| \le \delta$$

for all positive integers n. On dividing by 2^n and letting n approach infinity, we see that

$$l(x + y) = l(x) + l(y).$$

If we replace x by $2^n x$ in the inequality [2.1], we find that

$$\|f(2^n x)/2^n - f(x)\| \le \delta(1 - 2^{-n}).$$

Hence, in the limit, $\|l(x) - f(x)\| \le \delta$.

Suppose that there was another linear transformation $L(x)$ satisfying the inequality $\|L(x) - f(x)\| \leq \delta$ for all $x \in E$, and such that $l(y) \neq L(y)$ for some point y of E. For any integer $n > 2\delta/\|l(y) - L(y)\|$, it is obvious that $\|l(ny) - L(ny)\| > 2\delta$, which contradicts the inequalities $\|L(ny) - f(ny)\| \leq \delta$ and $\|l(ny) - f(ny)\| \leq \delta$. Hence, $l(x)$ is the unique linear solution of the inequality $\|l(x) - f(x)\| \leq \delta$. □

Next the character of the linear transformation $l(x)$ when continuity restrictions are placed on the transformation $f(x)$ is investigated.

Theorem 2.1.2 (Hyers [6]). *If under the hypotheses of Theorem 2.1.1 we suppose that $f(x)$ is continuous at a single point y of E, then $l(x)$ is continuous everywhere in E.*

Proof. Assume, contrary to the theorem, that the linear transformation $l(x)$ is not continuous. Then there exists an integer k and a sequence of points x_n of E converging to zero such that $\|l(x_n)\| > l/k$ for all positive integers n. Let m be an integer greater than $3k\delta$. Then

$$\|l(mx_n + y) - l(y)\| = \|l(mx_n)\| > 3\delta.$$

On the other hand,

$$\|l(mx_n + y) - l(y)\| \leq \|l(mx_n + y) - f(mx_n + y)\| + \|f(mx_n + y) - f(y)\| + \|f(y) - l(y)\| < 3\delta$$

for sufficiently large n, since $\lim_{n\to\infty} f(mx_n + y) = f(y)$. This contradiction establishes the theorem. □

Corollary 2.1.3 (Hyers [6]). *If under the hypotheses of Theorem 2.1.1 we suppose that for each $x \in E$ the function $f(tx)$ is a continuous function of the real variable t for $t \in \mathbb{R}$, then $l(x)$ is homogeneous of degree one.*

Proof. For fixed $x, f(tx)$ is a continuous δ-linear transformation of the real axis into E'. By Theorems 2.1.1 and 2.1.2, $l(tx)$ is continuous in t and hence, being linear, $l(x)$ is homogeneous of degree one. □

In 1991, Baker [7] used the Banach fixed point theorem to prove the Hyers-Ulam stability for single variable functional equations.

Proposition 2.1.4 (Baker [7]). *Suppose S is a nonempty set, $\phi : S \to S$, E is a Banach space, λ is a scalar, $|\lambda| < 1$, $\delta > 0$, and $f : S \to E$ such that*

$$\|\lambda f(\phi(x)) - f(x)\| \leq \delta$$

for all $x \in S$. Then there exists a unique $h : S \to E$ such that $\lambda h(\phi(x)) = h(x)$ for all $x \in S$ and

$$\|f(x) - h(x)\| \leq \delta(1 - |\lambda|)$$

for all $x \in S$.

Theorem 2.1.5 (Baker [7]). *Suppose (Y, ρ) is a complete metric space and $T : Y \to Y$ is a contraction (for some $\lambda \in [0, 1)$, $\rho(T(x), T(y)) < \lambda\rho(x, y)$ for all $x, y \in Y$). Also* *suppose that* $u \in Y$, $\delta > 0$ *and* $\rho(u, T(u)) < \delta$. *Then there exists a unique* $p \in Y$ *such that* $p = T(p)$. *Moreover,* $\rho(u, p) < \delta/(1 - \lambda)$.

Theorem 2.1.6 (Baker [7]). *Suppose S is a nonempty set;* (X, d) *is a complete metric space;* $\phi : S \to S$, $F : S \times X \to X$, $0 \leq \lambda < 1$; *and*

$$d(F(t, u), F(t, v)) \leq \lambda d(u, v)$$

for all $t \in S$ *and all* $u, v \in S$. *Also suppose that* $g : S \to X$, $\delta > 0$ *and*

$$d(g(t), F(t, g(\phi(t)))) \leq \delta \tag{2.2}$$

for all $t \in S$. *Then there is a unique function* $f : S \to X$ *such that*

$$f(t) = F(t, f(\phi(t))) \tag{2.3}$$

for all $t \in S$ *and*

$$d(f(t), g(t)) \leq \delta/(1 - \lambda) \tag{2.4}$$

for all $t \in S$.

Proof. Let $Y = \{a : S \to X | \sup\{d(a(t), g(t)) | t \in S\} < \infty\}$. For $a, b \in Y$ define $\rho(a, b) = \sup\{d(a(t), b(t)) | t \in S\}$. Then $g \in Y$, ρ is a metric on Y, and convergence with respect to ρ means uniform convergence on S with respect to d. Moreover, the completeness of X with respect to d implies the completeness of Y with respect to ρ.

For $a \in Y$, define $T(a) : S \to X$ by

$$(T(a))(t) = F(t, a(\phi(t)))$$

for all $t \in S$. Then T maps Y into Y. If $a, b \in Y$ then for all $t \in S$,

$$\begin{aligned} d((T(a))(t), (T(b))(t)) &= d(F(t, a(\phi(t))), F(t, b(\phi(t)))) \\ &\leq \lambda d(a(\phi(t)), b(\phi(t))) \\ &\leq \lambda \rho(a \circ \phi, b \circ \phi) \\ &\leq \lambda \rho(a, b). \end{aligned}$$

Thus,

$$\rho(T(a), T(b)) \leq \lambda \rho(a, b)$$

for all $a, b \in Y$. But Eq. [2.2] means that $\rho(g, T(g)) \leq \delta$. Hence, according to Theorem 2.1.5, there is a unique f in Y such that $f = T(f)$ and $\rho(g, f) \leq \delta/(1 - \lambda)$. That is, Eqs. [2.3] and [2.4] hold. □

2.2 Theorem of Themistocles M. Rassias

Theorem 2.2.1 (Th.M. Rassias [8]). *Consider E_1, E_2 to be two Banach spaces, and let $f : E_1 \to E_2$ be a mapping such that $f(tx)$ is continuous in t for each fixed x. Assume that there exists $\theta \geq 0$ and $0 \leq p < 1$ such that*

$$\frac{\|f(x+y) - f(x) - f(y)\|}{\|x\|^p + \|y\|^p} \leq \theta \quad \text{for any } x, y \in E_1.$$

Then there exists a unique linear mapping $T : E_1 \to E_2$ such that

$$\frac{\|f(x) - T(x)\|}{\|x\|^p} \leq \frac{2\theta}{2 - 2^p} \quad \text{for any } x \in E_1. \tag{2.5}$$

Proof. Claim that

$$\frac{\|f(2^n x)/2^n - f(x)\|}{\|x\|^p} \leq \theta \sum_{m=0}^{n-1} 2^{m(p-1)} \tag{2.6}$$

for any integer n, and some $\theta \geq 0$. The verification of Eq. [2.6] follows by induction on n. Indeed the case $n = 1$ is clear because by the hypothesis we can find θ, that is greater or equal to zero, and p such that $0 \leq p < 1$ with

$$\frac{\|f(2x)/2 - f(x)\|}{\|x\|^p} \leq \theta.$$

Assume now that Eq. [2.6] holds and we want to prove it for the case $(n+1)$. However, this is true because by Eq. [2.6] we obtain

$$\frac{\|f(2^n \cdot 2x)/2^n - f(2x)\|}{\|2x\|^p} \leq \theta \sum_{m=0}^{n-1} 2^{m(p-1)},$$

therefore,

$$\frac{\|f(2^{n+1}x)/2^{n+1} - 1/2f(2x)\|}{\|x\|^p} \leq \theta \sum_{m=1}^{n} 2^{m(p-1)}.$$

By the triangle inequality, we obtain

$$\left\|f(2^{n+1}x)/2^{n+1} - f(x)\right\| \leq \left\|f(2^{n+1}x)/2^{n+1} - 1/2f(2x)\right\| + \|1/2f(2x) - f(x)\|$$
$$\leq \theta \|x\|^p \sum_{m=0}^{n} 2^{m(p-1)}.$$

Thus,

$$\frac{\|f(2^{n+1}x)/2^{n+1} - f(x)\|}{\|x\|^p} \leq \theta \sum_{m=0}^{n} 2^{m(p-1)}$$

and Eq. [2.6] is valid for any integer n. It follows then that

$$\frac{\|f(2^n x)/2^n - f(x)\|}{\|x\|^p} \leq \frac{2\theta}{2 - 2^p}, \tag{2.7}$$

because $\sum_{m=0}^{n-1} 2^{m(p-1)}$ converges to $2/(2-2^p)$ as $0 \leq p < 1$. However, for $m > n > 0$,

$$\begin{aligned}\|f(2^m x)/2^m - f(2^n x)/2^n\| &= 1/2^n \|1/2^{m-n} f(2^m x) - f(2^n x)\| \\ &\leq \theta 2^{n(p-1)} \frac{2\theta}{2 - 2^p} \|x\|^p.\end{aligned}$$

Therefore,

$$\lim_{n\to\infty} \|f(2^m x)/2^m - f(2^n x)/2^n\| = 0.$$

But E_2, as a Banach space, is complete, thus the sequence $\{f(2^n x)/2^n\}$ converges. Set

$$T(x) = \lim_{n\to\infty} f(2^n x)/2^n.$$

It follows that

$$\begin{aligned}\|f(2^n(x+y)) - f(2^n x) - f(2^n y)\| &\leq \theta \left(\|2^n x\|^p + \|2^n y\|^p\right) \\ &= 2^{np}\theta \left(\|x\|^p + \|y\|^p\right).\end{aligned}$$

Therefore,

$$\frac{1}{2^n} \|f(2^n(x+y)) - f(2^n x) - f(2^n y)\| \leq 2^{n(p-1)}\theta \left(\|x\|^p + \|y\|^p\right)$$

or

$$\lim_{n\to\infty} \frac{1}{2^n} \|f(2^n(x+y)) - f(2^n x) - f(2^n y)\| \leq \lim_{n\to\infty} 2^{n(p-1)}\theta \left(\|x\|^p + \|y\|^p\right)$$

or

$$\left\| \lim_{n\to\infty} \frac{1}{2^n} f(2^n(x+y)) - \lim_{n\to\infty} \frac{1}{2^n} f(2^n x) - \lim_{n\to\infty} \frac{1}{2^n} f(2^n y) \right\| = 0$$

or

$$\|T(x+y) - T(x) - T(y)\| = 0 \quad \text{for any } x, y \in E_1$$

or

$$T(x+y) = T(x) + T(y)$$

for all $x, y \in E_1$. Since $T(x+y) = T(x) + T(y)$ for any $x, y \in E_1$, $T(rx) = rT(x)$ for any rational number r. Fix $x_0 \in E_1$ and $\rho \in E_2^*$ (the dual space of E_2). For $t \in \mathbb{R}$, consider the mapping $t \mapsto \rho(T(tx)) = \phi(t)$. Then $\phi : \mathbb{R} \to \mathbb{R}$ satisfies the property that $\phi(a+b) = \phi(a) + \phi(b)$, ie, ϕ is a group homomorphism. Moreover, ϕ is a Borel function, because of the following reasoning. Let $\phi(t) = \lim_{n\to\infty} \rho(f(2^n tx_0))/2^n$ and set $\phi_n(t) = \rho(f(2^n tx_0))/2^n$. Then $\phi_n(t)$ are continuous functions. But $\phi(t)$ is the pointwise limit of continuous functions, thus $\phi(t)$ is a Borel function. It is a known fact that if $\phi : \mathbb{R}^n \to \mathbb{R}^n$ is a function such that ϕ is a group homomorphism, ie, $\phi(x+y) = \phi(x) + \phi(y)$ and ϕ is a measurable function, then ϕ is continuous. In fact this statement is also true if we replace $\mathbb{R}^n$ by any separable, locally compact abelian group. Therefore, ϕ is a continuous function. Let $a \in \mathbb{R}$. Then $a = \lim_{n\to\infty} r_n$, where $\{r_n\}$ is a sequence of rational numbers. Hence,

$$\phi(at) = \phi(t \lim_{n\to\infty} r_n) = \lim_{n\to\infty} \phi(tr_n) = (\lim_{n\to\infty} r_n)\phi(t) = a\phi(t).$$

Therefore, $\phi(at) = a\phi(t)$ for any $a \in \mathbb{R}$. Thus $T(ax) = aT(x)$ for any $a \in \mathbb{R}$. Hence, T is a linear mapping.

From Eq. [2.7] we obtain

$$\frac{\|f(2^n x)/2^n - f(x)\|}{\|x\|^p} \le \frac{2\theta}{2-2^p},$$

or equivalently,

$$\frac{\|T(x) - f(x)\|}{\|x\|^p} \le \varepsilon, \tag{2.8}$$

where $\varepsilon = 2\theta/(2-2^p)$. Thus we have obtained Eq. [2.5].

We want now to prove that T is the unique such linear mapping. Assume that there exists another one, denoted by $g : E_1 \to E_2$ such that $T(x) \neq g(x)$ for some $x \in E_1$. Then there exists a constant ε_1, greater or equal to zero, and q such that $0 \le q < 1$ with

$$\frac{\|g(x) - f(x)\|}{\|x\|^q} \le \varepsilon_1.$$

By the triangle inequality and Eq. [2.8], we obtain

$$\|T(x) - g(x)\| \leq \|T(x) - f(x)\| + \|f(x) - g(x)\| \leq \varepsilon \|x\|^p + \varepsilon_1 \|x\|^q.$$

Therefore,

$$\|T(x) - g(x)\| = \left\| \frac{1}{n} T(nx) - \frac{1}{n} g(nx) \right\| = \frac{1}{n} \|T(nx) - g(nx)\|$$
$$\leq \frac{1}{n} (\varepsilon \|nx\|^p + \varepsilon_1 \|nx\|^q) = n^{p-1} \varepsilon \|x\|^p + n^{q-1} \varepsilon_1 \|x\|^q.$$

Thus $\lim_{n\to\infty} \|T(x) - g(x)\| = 0$ for all $x \in E_1$ and hence, $T(x) \equiv g(x)$ for all $x \in E_1$. □

In 1982, J.M. Rassias [9] considered the Cauchy difference controlled by a product of different powers of norm.

Theorem 2.2.2 (J.M. Rassias [9]). *Let A be a normed linear space with norm* $\|\cdot\|_1$ *and B be a Banach space with norm* $\|\cdot\|_2$. *Assume in addition that* $f : A \to B$ *is a mapping such that* $f(tx)$ *is continuous in t for each fixed x. If there exists* $0 \leq a < 1/2$ *and* $\delta > 0$ *such that*

$$\|f(x+y) - f(x) - f(y)\|_2 \leq 2\delta \|x\|_1^a \cdot \|y\|_1^a$$

for any $x, y \in A$, *then there exists a linear mapping* $L : A \to B$ *such that*

$$\|f(x) - L(x)\|_2 \leq c\|x\|_1^{2a}$$

for any $x \in A$, *where* $c = \delta/(1 - 2^{2a-1})$. *If* $M : A \to B$ *be a linear mapping such that*

$$\|f(x) - M(x)\|_2 \leq c' \|x\|_1^{2b}$$

for any $x \in A$, *where* $b : 0 \leq b < \frac{1}{2}$ *and* c' *is a constant, then* $L(x) = M(x)$ *for any* $x \in A$.

Gajda [10], following the same approach as in Rassias [8], obtained the result for the case $p > 1$. However, it was verified that the result for the case $p = 1$ does not hold.

Theorem 2.2.3 (Gajda [10]). *Let* E_1 *and* E_2 *be two (real) normed linear spaces and assume that* E_2 *is complete. Let* $f : E_1 \to E_2$ *be a mapping for which there exist two constants* $\varepsilon \in [0, \infty)$ *and* $p \in R \setminus \{1\}$ *such that*

$$\|f(x+y) - f(x) - f(y)\| \leq \varepsilon \|x\|^p + \|y\|^p$$

for all $x, y \in E_1$. *Then there exists a unique additive mapping* $T : E_1 \to E_2$ *such that*

$$\|f(x) - T(x)\| \leq \delta \|x\|^p$$

for all $x \in E_1$, *where*

$$\delta = \begin{cases} \frac{2\varepsilon}{2-2^p}, & p < 1; \\ \frac{2\varepsilon}{2^p-2}, & p > 1. \end{cases}$$

Moreover, if for each $x \in E_1$ *the transformation* $t \mapsto f(tx)$ *is continuous for all* $t \in \mathbb{R}$, *then the mapping* T *is linear.*

Rassias and Šemrl [11] generalized the results of Gajda as follows.

Theorem 2.2.4 (Rassias and Šemrl [11]). *Let* $\beta(s,t)$ *be nonnegative for all nonnegative real numbers* s, t *and positive homogeneous of degree* p, *where* p *is real and* $p \neq 1$, *ie,* $\beta(\lambda s, \lambda t) = \lambda^p \beta(s,t)$ *for all nonnegative* λ, s, t. *Given a normed space* E_1 *and a Banach space* E_2, *assume that* $f : E_1 \to E_2$ *satisfies the inequality*

$$\|f(x+y) - f(x) - f(y)\| \leq \beta(\|x\|, \|y\|)$$

for all $x, y \in E_1$, *where*

$$\delta := \begin{cases} \frac{\beta(1,1)}{2-2^p}, & \text{if } p < 1; \\ \frac{\beta(1,1)}{2-2^p}, & \text{if } p > 1. \end{cases}$$

Corollary 2.2.5 (Rassias and Šemrl [11]). *Let* $f : E_1 \to E_2$ *be a mapping satisfying the hypotheses of Theorem 2.2.4 and suppose that* f *is continuous at a single point* $y \in E_1$. *Then the additive mapping* g *is continuous.*

Corollary 2.2.6 (Rassias and Šemrl [11]). *If, under the hypotheses of Theorem 2.2.4, we assume that, for each fixed* $x \in E_1$, *the mapping* $t \to f(tx)$ *from* $\mathbb{R}$ *to* E_2 *is continuous, then the additive mapping* g *is linear.*

A further generalization of the Hyers-Ulam stability for a large class of mappings was obtained by Isac and Rassias. They also presented some applications in nonlinear analysis, especially in fixed point theory.

Theorem 2.2.7 (Isac and Rassias [12]). *Let* E_1 *and* E_2 *be a real normed space and a real Banach space, respectively. Let* $\psi : [0,\infty) \to [0,\infty)$ *be a function satisfying the following conditions:*

(R_1) $\lim_{t\to\infty} \psi(t)/t = 0$,
(R_2) $\psi(ts) \leq \psi(t)\psi(s)$ *for all* $t, s \in [0,\infty)$,
(R_3) $\psi(t) \leq t$ *for all* $t > 1$.

If a function $f : E_1 \to E_2$ *satisfies the inequality*

$$\|f(x+y) - f(x) - f(y)\| \leq \theta\,(\psi(\|x\|) + \psi(\|y\|))$$

for some $\theta \geq 0$ *and for all* $x, y \in E_1$, *then there exists a unique additive function* $A : E_1 \to E_2$ *such that*

$$\|f(x) - A(x)\| \leq \frac{2\theta}{2 - \psi(2)} \psi(\|x\|)$$

for all $x \in E_1$. Moreover, if $f(tx)$ is continuous in t for each fixed x, then the function A is linear.

The method of Baker in Section 2.1 was generalized in 2003 by Radu [13], as follows.

Theorem 2.2.8 (Banach contraction principle). *Let (X, m) be a complete generalized metric space and consider a mapping $T : X \to X$ as a strictly contractive mapping, that is*

$$m(Tx, Ty) \leq Lm(x, y)$$

for all $x, y \in X$ and for some (Lipschitz constant) $0 < L < 1$. Then

- *T has one and only one fixed point $x^* = T(x^*)$;*
- *x^* is globally attractive, that is, $\lim_{n\to\infty} T^n x = x^*$ for any starting point $x \in X$; and*
- *one has the following estimation inequalities for all $x \in X$ and $n \geq 0$*

$$m(T^n x, x^*) \leq L^n m(x, x^*),$$

$$m(T^n x, x^*) \leq \frac{1}{1-L} L^n m(T^n x, T^{n+1} x),$$

$$m(x, x^*) \leq \frac{1}{1-L} m(x, Tx).$$

Theorem 2.2.9 (The Alternative of Fixed Point [14]). *Suppose that we are given a complete generalized metric space (X, m) and a strictly contractive mapping $T : X \to X$ with Lipschitz constant L. Then, for each given element $x \in X$, either $m(T^n x, T^{n+1} x) = +\infty$ for all nonnegative integers n or there exists a positive integer n_0 such that $m(T^n x, T^{n+1} x) < +\infty$ for all $n \geq n_0$. If the second alternative holds, then*

- ⋆ *the sequence $(T^n x)$ is convergent to a fixed point y^* of T;*
- ⋆ *y^* is the unique fixed point of T in the set $Y = \{y \in X, m(T^{n_0} x, y) < +\infty\}$; and*
- ⋆ *$m(y, y^*) \leq \frac{1}{1-L} m(y, Ty)$, $y \in Y$.*

Radu showed that the theorems of Hyers, Rassias, and Gajda concerning the stability of Cauchy's functional equation in Banach spaces are direct consequences of the alternative of fixed point.

2.3 Completeness of normed spaces

In most stability theorems for functional equations, the completeness of the target space of the unknown functions contained in the equation is assumed. In this section, we investigate the question of whether the stability of a functional equation implies this completeness.

During the 25th International Symposium on Functional Equations in 1988, this problem was considered by Schwaiger, who proved that if X is a normed space, then the stability of Cauchy's functional equation $f(x+y) = f(x) + f(y)$ $(x, y \in \mathbb{Z})$ for functions $f : \mathbb{Z} \to X$ implies the completeness of X (cf. [15]). Forti and Schwaiger [16] proved that an analogous statement is valid if the domain of f is an abelian group containing an element of infinite order. Moszner [17] showed that the assumptions of this theorem are essential and presented some applications of the result above.

Let $(G, +)$ be an abelian group and let E be a normed space. A mapping $f : G \to E$ is called ε-quadratic if for a given $\varepsilon > 0$ it satisfies

$$\|f(x+y) + f(x-y) - 2f(x) - 2f(y)\| \leq \varepsilon$$

for all $x, y \in G$. In 2010, Najati [18] showed that E is complete if every ε-quadratic mapping $f : X \to E$ can be estimated by a quadratic mapping, where X is $\mathbb{N}_0$ or a finitely generated free abelian group. In this section, we consider the results given by Najati [18].

Let us denote the sets $\underbrace{\mathbb{N}_0 \times \cdots \times \mathbb{N}_0}_{r\text{-times}}$ and $\underbrace{\mathbb{Z} \times \cdots \times \mathbb{Z}}_{r\text{-times}}$ by $\mathbb{N}_0^r$ and $\mathbb{Z}^r$, respectively. The proof of the following lemma is obvious:

Lemma 2.3.1 (Najati [18]). *Let $f : \mathbb{N}_0^r \to E$ ($f : \mathbb{Z}^r \to E$) be a quadratic mapping, ie,*

$$f(m_1 + n_1, \ldots, m_r + n_r) + f(m_1 - n_1, \ldots, m_r - n_r) = 2f(m_1, \ldots, m_r) + 2f(n_1, \ldots, n_r)$$

for all $m_i \geq n_i \geq 0$ ($m_i, n_i \in \mathbb{Z}$) with $1 \leq i \leq r$. Then

$$f(m, \ldots, m) = m^2 f(1, \ldots, 1)$$

for all $m \in \mathbb{N}_0$ ($m \in \mathbb{Z}$).

Lemma 2.3.2 (Najati [18]). *Let $f : \mathbb{N}_0 \to E$ be an ε-quadratic mapping, ie,*

$$\|f(m+n) + f(m-n) - 2f(m) - 2f(n)\| \leq \varepsilon \quad (m \geq n \geq 0).$$

Then the mapping $g : \mathbb{Z} \to E$ defined by $g(n) = f(|n|)$ is ε-quadratic.

Proof. For convenience, we use the following abbreviations:

$$Qf(m,n) = f(m+n) + f(m-n) - 2f(m) - 2f(n) \quad (m \geq n \geq 0),$$
$$Qg(m,n) = g(m+n) + g(m-n) - 2g(m) - 2g(n) \quad m, n \in \mathbb{Z}.$$

It follows from the definition of g that

$$Qf(m,n)=\begin{cases} Qf(m,n), & \text{if } m\geq n\geq 0,\\ Qf(n,m), & \text{if } m,n\geq 0,\ m<n,\\ Qf(-n,-m), & \text{if } m,n<0,\ m\geq n,\\ Qf(-m,-n), & \text{if } m,n<0,\ m<n,\\ Qg(m,-n), & \text{if } n<0\leq m,\ m+n\geq 0,\\ Qg(-n,m), & \text{if } n<0\leq m,\ m+n<0,\\ Qg(n,-m), & \text{if } m<0\leq n,\ m+n\geq 0,\\ Qg(-m,n), & \text{if } m<0\leq n,\ m+n<0. \end{cases}$$

Indeed,

$$Qf(m,n)=\begin{cases} Qf(|m|,|n|), & \text{if } |m|\geq |n|,\\ Qf(|n|,|m|), & \text{if } |n|\geq |m|, \end{cases}$$

for all $m,n\in\mathbb{Z}$. Therefore, g is ε-quadratic on $\mathbb{Z}$. □

Using the proof of Lemma 2.3.2, we have the following proposition.

Proposition 2.3.3 (Najati [18]). *Let $f:\mathbb{N}_0^r\to E$ be an ε-quadratic mapping. Then the mapping $g:\mathbb{Z}^r\to E$ defined by*

$$g(n_1,n_2,\ldots,n_r)=f(|n_1|,|n_2|,\ldots,|n_r|)$$

is ε-quadratic.

Theorem 2.3.4 (Najati [18]). *Let E be a normed space such that for each ε-quadratic mapping $f:\mathbb{Z}^r\to E$ there exists a quadratic mapping $Q:\mathbb{Z}^r\to E$ such that $Q-f$ is uniformly bounded on $\mathbb{Z}^r$. Then E is complete.*

Proof. Let us have $\varepsilon>0$ and let $\{x_n\}_{n\geq 1}$ be a Cauchy sequence in E. There exists a subsequence $\{x_{n_k}\}_{k\geq 1}$ such that $\|X_n-x_m\|<\frac{\varepsilon}{6k^2}$ for all $m,n\geq n_k$. Let $y_0=0$ and $y_k=x_{n_k}$ for all $k\geq 1$. We define the mapping $f:\mathbb{N}_0^r\to E$ by $f(k_1,\ldots,k_r)=\frac{1}{r}\sum_{i=1}^r k_I^2 y_{k_i}$ for all $(k_1,\ldots,k_r)\in\mathbb{N}_0^r$. Then

$$\begin{aligned}
&\|f(m_1+k_1,\ldots,m_r+k_r)+f(m_1-k_1,\ldots,m_r-k_r)-2f(m_1,\ldots,m_r)\\
&-2f(k_1,\ldots,k_r)\|\\
&=\frac{1}{r}\left\|\sum_{i=1}^r[(m_i^2+2m_ik_i)(y_{m_i+k_i}-y_{m_i})+k_i^2(y_{m_i+k_i}-y_{k_i})\right.\\
&\left.\quad+(m_i-k_i)^2(y_{m_i-k_i}-y_{m_i})+k_i^2(y_{m_i}-y_{k_i})]\right\|\\
&\leq\frac{1}{r}\sum_{i=1}^r[(m_i^2+2m_ik_i)\|y_{m_i+k_i}-y_{m_i}\|+k_i^2\|y_{m_i+k_i}-y_{k_i}\|\\
&\quad+(m_i-k_i)^2\|y_{m_i-k_i}-y_{m_i}\|+k_i^2\|y_{m_i}-y_{k_i}\|]\\
&\leq\frac{1}{r}\sum_{i=1}^r\left[\frac{\varepsilon}{6}+\frac{2k_i\varepsilon}{6m_i}+\frac{\varepsilon}{6}+\frac{\varepsilon}{6}+\frac{\varepsilon}{6}\right]\leq\varepsilon
\end{aligned}$$

for all $m_i > k_i \geq 0$, where $1 \leq i \leq r$. Hence, it follows from $f(0,\ldots,0) = 0$ that f is ε-quadratic. By Lemma 2.3.2, the mapping $g : \mathbb{Z}^r \to E$ defined by $g(k_1,\ldots,k_r) = f(|k_1|,\ldots,|k_r|)$ is ε-quadratic. By our assumption, there exists a quadratic mapping $Q : \mathbb{Z}^r \to E$ and a positive constant M such that

$$\|g(k_1,\ldots,k_r) - Q(k_1,\ldots,k_r)\| \leq M,$$

$(k_1,\ldots,k_r) \in \mathbb{Z}^r$. Since Q is quadratic on $\mathbb{Z}^r$, by Lemma 2.3.1 we have $Q(k,\ldots,k) = k^2Q(1,\ldots,1)$ for all $k \in \mathbb{Z}$. Therefore,

$$\|k^2x_{n_k} - k^2Q(1,\ldots,1)\| \leq M$$

for all positive integers k. This shows that the subsequence $\{x_{n_k}\}_{k\geq 1}$ converges to $Q(1,\ldots,1)$. Hence, the Cauchy sequence $\{x_n\}_{n\geq 1}$ also converges to $Q(1,\ldots,1)$, and the theorem is proved. □

The following result follows from the proof of Theorem 2.3.4.

Theorem 2.3.5 (Najati [18]). *Let E be a normed space such that for each ε-quadratic mapping $f : \mathbb{N}_0^r \to E$ there exists a quadratic mapping $Q : \mathbb{N}_0^r \to E$ such that $Q - f$ is uniformly bounded on $\mathbb{N}_0^r$. Then E is complete.*

Corollary 2.3.6 (Najati [18]). *Let E be a normed space such that for each ε-quadratic mapping $f : \mathbb{Z} \to E$ ($f : \mathbb{N}_0 \to E$) there exists a quadratic mapping $Q : \mathbb{Z} \to E$ ($Q : \mathbb{N}_0 \to E$) such that $Q - f$ is uniformly bounded on $\mathbb{Z}$ ($\mathbb{N}_0$). Then E is complete.*

Theorem 2.3.4 remains true if we replace $\mathbb{Z}^r$ by a finitely generated free abelian group.

Theorem 2.3.7 (Najati [18]). *Let G be a finitely generated free abelian group and E be a normed space such that for each ε-quadratic mapping $f : G \to E$ there exists a quadratic mapping $Q : G \to E$ such that $Q - f$ is uniformly bounded on G. Then E is complete.*

Proof. Let $\{g_1,\ldots,g_r\}$ be a basis for G and let $f : \mathbb{Z}^r \to E$ be an ε-quadratic mapping. We define the mapping $\phi : G \to E$ by $\phi\left(\sum_{i=1}^r n_ig_i\right) = f(n_1,\ldots,n_r)$. It is clear that ϕ is ε-quadratic. By the assumption, there exists a quadratic mapping $Q : G \to E$ such that $Q - \phi$ is uniformly bounded on G. Let $T : \mathbb{Z}^r \to E$ be a mapping defined by $T(n_1,\ldots,n_r) = Q\left(\sum_{i=1}^r n_ig_i\right)$. Hence, T is quadratic and $T - f$ is uniformly bounded on $\mathbb{Z}^r$. Hence, by Theorem 2.3.4, E is complete. □

In 2013, Fošner et al. [19] gave a type of characterization of Banach spaces in terms of the stability of functional equations. More precisely, they proved that a normed space X is complete if there exists a functional equation of the type

$$\sum_{i=1}^{n} a_if(\phi_i(x_1,\ldots,x_k)) = 0 \quad (x_1,\ldots,x_k \in D) \tag{2.9}$$

with given real numbers $a_1,\ldots,a_n$, given mappings $\phi_1,\ldots,\phi_n : D^k \to D$ and unknown function $f : D \to X$, which has a Hyers-Ulam stability property on an infinite subset D of the integers.

We say that the functional equation [2.9] satisfies a (strong) Hyers-Ulam stability property on D if there exists a polynomial p of degree of at least 1 on D such that pb solves Eq. [2.9] with some $b \in X$, $b \neq 0$; furthermore, if, for a function $f : D \to X$, the left-hand side of Eq. [2.9] is bounded, then there exists a function $g : D \to X$, which has the form $g(x) = p(x)c$, $(x \in D)$, with a $c \in X$, satisfies the Eq. [2.9] and the difference $f - g$ is uniformly bounded on D.

The class of equations given in Eq. [2.9] contains the Cauchy equation

$$f(x+y) = f(x) + f(y) \quad (x, y \in \mathbb{Z}),$$

the square norm equation

$$f(x+y) + f(x-y) = 2f(x) + 2f(y) \quad (x, y \in \mathbb{Z}),$$

and several other well-known functional equations as a special case. It is also easy to see that monomial functional equations, polynomial functional equations and, using the notation above, even linear functional equations of the form

$$\sum_{i=1}^{n} a_i f(p_i x + q_i y) = 0 \quad (x, y \in \mathbb{Z}), \tag{2.10}$$

where $p_1, \ldots, p_n$ and $q_1, \ldots, q_n$ are given integers and $f : \mathbb{Z} \to X$ is an unknown function, are contained in the class [2.9]. In the next theorem, a reverse statement of Hyers-Ulam stability is considered.

Theorem 2.3.8 (Fošner et al. [19]). *Let X be a normed space. If there exists a functional equation of type* [2.9] *which has the Hyers-Ulam stability property above on an infinite subset D of the integers, then X is complete.*

Proof. Let $\bar{X}$ be the completion of X and let $\bar{c} \in \bar{X}$. Let, furthermore, $g : D \to \bar{X}$ be given by the formula $g(x) = p(x)\bar{c}$, $(x \in D)$, where p is the polynomial considered in the definition of the stability property. Since X is dense in $\bar{X}$, there exists a function $h : D \to X$ such that

$$\|g(x) - h(x)\| \leq 1 \quad (x \in D). \tag{2.11}$$

Obviously, g is a solution of Eq. [2.9]; therefore,

$$\begin{aligned}
&\left\| \sum_{i=1}^{n} a_i h(\phi_i(x_1, \ldots, x_k)) \right\| \\
&\quad \leq \left\| \sum_{i=1}^{n} a_i h(\phi_i(x_1, \ldots, x_k)) - \sum_{i=1}^{n} a_i g\big(\phi_i(x_1, \ldots, x_k)\big) \right\| \\
&\quad\quad + \left\| \sum_{i=1}^{n} a_i g\big(\phi_i(x_1, \ldots, x_k)\big) \right\| \leq \sum_{i=1}^{n} |a_i| = K \quad (x_1, \ldots, x_k \in D).
\end{aligned}$$

The stability property implies the existence of a $c \in X\,\{0\}$ such that

$$\|h(x) - p(x)c\| \leq L \quad (x \in D) \tag{2.12}$$

with some $L \in \mathbb{R}$. Using the inequalities [2.11] and [2.12], we obtain

$$\|p(x)\bar{c} - p(x)c\| \leq \|p(x)\bar{c} - h(x)\| + \|h(x) - p(x)c\| \leq 1 + L,$$

thus,

$$\|\bar{c} - c\| \leq \frac{1+L}{|p(x)|} \quad (x \in D, p(x) \neq 0).$$

Letting $|x| \to \infty$, we get that $\bar{c} = c$, which yields our statement. □

Stability of mixed type functional equations

3

3.1 Binary mixtures of functional equations

The functional equation

$$f(x+y)+f(x-y)=2f(x)+2f(y) \tag{3.1}$$

is related to a symmetric bi-additive function [20, 21]. The functional equation [3.1] is called quadratic. In particular, every solution of the quadratic equation [3.1] is said to be a quadratic function. It is well known that a function f between real vector spaces is quadratic if and only if there exists a unique symmetric bi-additive function B such that $f(x) = B(x,x)$ for all x (see [20, 21]. The bi-additive function B is given by

$$B(x,y)=\frac{1}{4}[f(x+y)-f(x-y)].$$

Jun and Kim [22] introduced the following cubic functional equation:

$$f(2x+y)+f(2x-y)=2f(x+y)+2f(x-y)+12f(x) \tag{3.2}$$

and they established the general solution and the generalized Hyers-Ulam stability problem for the functional equation [3.2]. Let E_1 and E_2 be real vector spaces. They proved that a function $f : E_1 \to E_2$ satisfies the functional equation [3.2] if and only if there exists a function $G : E_1 \times E_1 \times E_1 \to E_2$ such that $f(x) = G(x,x,x)$ for all $x \in E_1$, and G is symmetric for each fixed one variable and additive for each fixed two variables. The function G is given by

$$G(x,y,z)=\frac{1}{24}[f(x+y+z)+f(x-y-z)-f(x+y-z)-f(x-y+z)]$$

for all $x, y, z \in E_1$. It is easy to see that the function $f(x) = cx^3$ is a solution of the functional equation [3.2]. Thus, it is natural that Eq. [3.2] is called a cubic functional equation and every solution of the cubic functional equation [3.2] is said to be a cubic function.

Theory of Approximate Functional Equations. http://dx.doi.org/10.1016/B978-0-12-803920-5.00003-1

Chang and Jung [23] introduced the following functional equation deriving from cubic and quadratic functions:

$$6f(x+y) - 6f(x-y) + 4f(3y) = 3f(x+2y) - 3f(x-2y) + 9f(2y). \tag{3.3}$$

It is easy to see that the function $f(x) = ax^2 + bx^3$ is a solution of Eq. [3.3]. We deal with the solution and stability of the functional equation [3.3].

Lemma 3.1.1 (Chang and Jung [23]). *Let X and Y be real vector spaces. If an even function $f : X \to Y$ satisfies Eq. [3.3] for all $x, y \in X$, then f is quadratic.*

Lemma 3.1.2 (Chang and Jung [23]). *Let X and Y be real vector spaces. If an odd function $f : X \to Y$ satisfies Eq. [3.3] for all $x, y \in X$, then f is cubic.*

Theorem 3.1.3 (Chang and Jung [23]). *A function $f : X \to Y$ satisfies Eq. [3.3] for all $x, y \in X$ if and only if there exist a symmetric biadditive function $B : X \times X \to Y$ and a function $G : X \times X \times X \to Y$ such that $f(x) = G(x, x, x) + B(x, x)$ for all $x \in X$, and G is symmetric for fixed one variable and G is additive for fixed two variables.*

Proof. If there exists a symmetric biadditive function $B : X \times X \to Y$ and a function $G : X \times X \times X \to Y$ such that $f(x) = G(x, x, x) + B(x, x)$ for all $x \in X$, and G is symmetric for fixed one variable and C is additive for fixed two variables, it is obvious that f satisfies Eq. [3.3].

Conversely, we decompose f into the even part and the odd part by putting

$$f_e(x) = \frac{1}{2}[f(x) + f(-x)] \quad \text{and} \quad f_o(x) = \frac{1}{2}[f(x) - f(-x)]$$

for all $x \in X$. It is easy to show by Lemmas 3.1.1 and 3.1.2 that we achieve the result. □

Now, we prove the stability of Eq. [3.3] in the spirit of generalized Hyers-Ulam. For convenience, we use the following abbreviation:

$$Df(x,y) := 6f(x+y) - 6f(x-y) + 4f(3y) - 3f(x+2y) + 3f(x-2y) - 9f(2y).$$

Theorem 3.1.4 (Chang and Jung [23]). *Let X be a real vector space and Y be a real Banach space. Let $\phi : X \times X \to [0, \infty)$ be a function such that*

$$\sum_{i=0}^{\infty} \frac{\phi(0, 2^i x) + 4\phi(2^i x, 2^i x)}{4^i}$$

converges and

$$\lim_{n\to\infty} \frac{\phi(2^n x, 2^n y)}{4^n} = 0$$

for all $x, y \in X$ and $f(0) = 0$. Suppose that an even function $f : X \to Y$ satisfies the inequality

$$\|Df(x,y)\| \leq \phi(x,y) \tag{3.4}$$

for all $x, y \in X$. Then the limit

$$Q(x) = \lim_{n\to\infty} \frac{f(2^n x)}{4^n} \tag{3.5}$$

exists for all $x \in X$ and $Q : X \to Y$ is a unique quadratic function satisfying Eq. [3.3] and

$$\|f(x) - Q(x)\| \leq \frac{1}{12}\sum_{i=0}^{\infty} \frac{\phi(0, 2^i x) + 4\phi(2^i x, 2^i x)}{4^i} \tag{3.6}$$

for all $x \in X$.

Proof. In Eq. [3.4], we set $x = 0$, then replace y by x, and get

$$\|4f(3x) - 9f(2x)\| \leq \phi(0, x) \tag{3.7}$$

for all $x \in X$. Put $y = x$ in Eq. [3.4] to obtain

$$\|f(3x) + 3f(x) - 3f(2x)\| \leq \phi(x, x) \tag{3.8}$$

for all $x \in X$. It follows from Eqs. [3.7] and [3.8] that

$$\begin{aligned}\|3f(2x) - 12f(x)\| &\leq 4\|f(3x) + 3f(x) - 3f(2x)\| + \|4f(3x) - 9f(2x)\| \\ &\leq 4\phi(x,x) + \phi(0,x)\end{aligned}$$

for all $x \in X$. Dividing by 12 in the above inequality, we have

$$\left\|\frac{f(2x)}{4} - f(x)\right\| \leq \frac{1}{12}[4\phi(x,x) + \phi(0,x)] \tag{3.9}$$

for all $x \in X$. Let us replace x by $2x$ in Eq. [3.9] and then divide by 4. Then the resulting inequality with Eq. [3.9] gives

$$\left\|\frac{f(2^2 x)}{4^2} - f(x)\right\| \leq \frac{1}{12}\left(\left[\frac{4\phi(2x,2x) + \phi(0,2x)}{4}\right] + [4\phi(x,x) + \phi(0,x)]\right) \tag{3.10}$$

for all $x \in X$. An induction argument now implies

$$\|4^{-n} f(2^n x) - f(x)\| \leq \frac{1}{12}\sum_{i=0}^{n-1} \frac{4\phi(2^i x, 2^i x) + \phi(0, 2^i x)}{4^i} \tag{3.11}$$

for all $x \in X$. We divide Eq. [3.11] by 4^m and replace x by $2^m x$ to obtain that

$$\begin{aligned}\|4^{-(n+m)}f(2^n2^mx) - 4^{-m}f(2^mx)\| &= 4^{-m}\|4^{-n}f(2^n2^mx) - f(2^mx)\| \\ &\le \frac{1}{12\cdot 4^m}\sum_{i=0}^{n-1}\frac{4\phi(2^i2^mx, 2^i2^mx) + \phi(0, 2^i2^mx)}{4^i} \\ &\le \frac{1}{12}\sum_{i=0}^{\infty}\frac{4\phi(2^i2^mx, 2^i2^mx) + \phi(0, 2^i2^mx)}{4^{m+i}}\end{aligned} \tag{3.12}$$

for all $x \in X$. This shows that $\{4^{-n}f(2^nx)\}$ is a Cauchy sequence in X by taking the limit $m \to \infty$. Since Y is a Banach space, it follows that the sequence $\{4^{-n}f(2^nx)\}$ converges.

We define $Q : X \to Y$ by $Q(x) = \lim_{n\to\infty}\frac{f(2^nx)}{4^n}$ for all $x \in X$. It is clear that $Q(-x) = Q(x)$ for all $x \in X$, and it follows from Eq. [3.4] that

$$\|DQ(x,y)\| \le \lim_{n\to\infty} 4^{-n}\|Df(2^nx, 2^ny)\| \le \lim_{n\to\infty} 4^{-n}\phi(2^nx, 2^ny) = 0$$

for all $x, y \in X$. Hence by Lemma 3.1.1, Q is quadratic.

It remains to show that Q is unique. Suppose that there exists another quadratic function $\tilde{Q} : X \to Y$ which satisfies Eqs. [3.3] and [3.6]. Since $\tilde{Q}(2^nx) = 4^n\tilde{Q}(x)$ and $Q(2^nx) = 4^nQ(x)$ for all $x \in X$, we conclude that

$$\begin{aligned}\|\tilde{Q}(x) - Q(x)\| &= 4^{-n}\|\tilde{Q}(2^nx) - Q(2^nx)\| \\ &\le 4^{-n}(\|\tilde{Q}(2^nx) - f(2^nx)\| + \|f(2^nx) - Q(2^nx)\|) \\ &\le \frac{1}{6}\sum_{i=0}^{\infty}\frac{\phi(0, 2^i2^nx) + 4\phi(2^i2^nx, 2^i2^nx)}{4^{n+i}}\end{aligned}$$

for all $x \in X$. By letting $n \to \infty$ in this inequality, we have $\tilde{Q}(x) = Q(x)$ for all $x \in X$, which gives the conclusion. □

Theorem 3.1.5 (Chang and Jung [23]). *Let X be a real vector space and Y be a real Banach space. Let $\phi : X \times X \to [0, \infty)$ be a function such that*

$$\sum_{i=0}^{\infty}\frac{\phi(0, 2^ix) + 4\phi(2^ix, 2^ix)}{8^i}$$

converges and

$$\lim_{n\to\infty}\frac{\phi(2^nx, 2^ny)}{8^n} = 0$$

for all $x, y \in X$. Suppose that an odd function $f : X \to Y$ satisfies the inequality

$$\|Df(x,y)\| \le \phi(x,y) \tag{3.13}$$

for all $x, y \in X$. Then the limit

$$C(x) = \lim_{n \to \infty} \frac{f(2^n x)}{8^n} \tag{3.14}$$

exists for all $x \in X$ and $C : X \to Y$ is a unique quadratic function satisfying Eq. [3.3] and

$$\|f(x) - C(x)\| \leq \frac{1}{24} \sum_{i=0}^{\infty} \frac{\phi(0, 2^i x) + 4\phi(2^i x, 2^i x)}{8^i} \tag{3.15}$$

for all $x \in X$.

Proof. Note that $f(x) = f(x)$ for all $x \in X$ and $f(0) = 0$ since f is odd. For $x = 0$, Eq. [3.13] implies that

$$\|12f(y) + 4f(3y) - 15f(2y)\| \leq \phi(0, y),$$

which on replacing y by x yields

$$\|12f(x) + 4f(3x) - 15f(2x)\| \leq \phi(0, x) \tag{3.16}$$

for all $x \in X$. If we put $y = x$ in Eq. [3.3], we get

$$\|3f(x) - f(3x) + 3f(2x)\| \leq \phi(x, x) \tag{3.17}$$

for all $x \in X$. According to Eqs. [3.16] and [3.17], we have

$$\begin{aligned} \|24f(x) - 3f(2x)\| &\leq \|12f(x) + 4f(3x) - 15f(2x)\| + 4\|3f(x) - f(3x) + 3f(2x)\| \\ &\leq \phi(0, x) + 4\phi(x, x) \end{aligned}$$

for all $x \in X$. By multiplying by $1/24$ in the above inequality, we obtain

$$\left\| \frac{f(2x)}{8} - f(x) \right\| \leq \frac{1}{24} [\phi(0, x) + 4\phi(x, x)] \tag{3.18}$$

for all $x \in X$. In Eq. [3.18], let us put $2x$ instead of x and then multiply by $1/8$. Then we have

$$\left\| \frac{f(2^2 x)}{8^2} - f(x) \right\| \leq \frac{1}{24} \left[\frac{\phi(0, 2x)}{8} + \frac{\phi(2x, 2x)}{2} + \phi(0, x) + 4\phi(x, x) \right] \tag{3.19}$$

for all $x \in X$. Applying an induction argument to n, we obtain

$$\|8^{-n} f(2^n x) - f(x)\| \leq \frac{1}{24} \sum_{i=0}^{n-1} \frac{\phi(0, 2^i x) + 4\phi(2^i x, 2^i x)}{8^i} \tag{3.20}$$

for all $x \in X$. Dividing Eq. [3.20] by 8^m and then substituting x by $2^m x$, we get

$$\begin{aligned}\|8^{-(n+m)}f(2^n2^mx) - 8^{-m}f(2^mx)\| &= 8^{-m}\|8^{-n}f(2^n2^mx) - f(2^mx)\| \\ &\leq \frac{1}{24\cdot 8^m}\sum_{i=0}^{n-1}\frac{\phi(0,2^i2^mx)+4\phi(2^i2^mx,2^i2^mx)}{8^i} \\ &\leq \frac{1}{24}\sum_{i=0}^{\infty}\frac{\phi(0,2^i2^mx)+4\phi(2^i2^mx,2^i2^mx)}{8^{m+i}}\end{aligned} \tag{3.21}$$

for all $x \in X$. Since the right-hand side of Eq. [3.21] tends to zero as $m \to \infty$, $\{8^{-n}f(2^nx)\}$ is a Cauchy sequence in Y and so converges. Define $C : X \to Y$ by $C(x) = \lim_{n\to\infty} 8^{-n}f(2^nx)$ for all $x \in X$. Since $C(-x) = -C(x)$ for all $x \in X$ and $\|DC(x,y)\| = 0$ for all $x, y \in X$ as in the proof of Theorem 3.1.4, and so from Lemma 3.1.2, it follows that C is cubic.

We claim that C is unique. Let $\tilde{C} : X \to Y$ be another cubic function satisfying Eqs. [3.3] and [3.15]. Since $\tilde{C}(2^nx) = 8^n\tilde{C}(x)$ and $C(2^nx) = 8^nC(x)$ for all $x \in X$, the rest of the proof is similar to the corresponding part of the proof of Theorem 3.1.4 and completes the proof. □

Theorem 3.1.6 (Chang and Jung [23]). *Let X be a real vector space and Y be a real Banach space. Let $\phi : X \times X \to [0,\infty)$ be a function such that*

$$\sum_{i=0}^{\infty}\frac{\phi(0,2^ix)+4\phi(2^ix,2^ix)}{4^i}$$

converges and

$$\lim_{n\to\infty}\frac{\phi(2^nx,2^ny)}{4^n} = 0$$

for all $x, y \in X$. Suppose that an odd function $f : X \to Y$ satisfies the inequality

$$\|Df(x,y)\| \leq \phi(x,y) \tag{3.22}$$

for all $x, y \in X$. Then there exists a unique cubic function $C : X \to Y$ and a unique quadratic function $Q : X \to Y$ satisfying Eq. [3.3] and

$$\|f(x) - C(x) - Q(x)\| \leq \frac{1}{24}\sum_{i=0}^{\infty}\left[\frac{\phi(0,2^ix)+4\phi(2^ix,2^ix)}{4^i} + \frac{\phi(0,-2^ix)+4\phi(-2^ix,-2^ix)+\phi(0,2^ix)+4\phi(2^ix,2^ix)+\phi(0,2^ix)+4\phi(2^ix,2^ix)}{2\cdot 8^i}\right] \tag{3.23}$$

for all $x \in X$.

Proof. Let $f_e(x) = \frac{1}{2}(f(x) + f(-x))$ for all $x \in X$. Then $f_e(0) = 0, f_e(-x) = f_e(x)$, and

$$\|Df_e(x, y)\| \leq \frac{1}{2}[\phi(x, y) + \phi(-x, -y)]$$

for all $x, y \in X$. Hence, in view of Theorem 3.1.4, there exists a unique quadratic function $q : X \to Y$ satisfying Eq. [3.5]. Let $f_o(x) = \frac{1}{2}(f(x) - f(-x))$ for all $x \in X$. Then $f_o(0) = 0, f_o(-x) = -f_o(x)$ and

$$\|Df_o(x, y)\| \leq \frac{1}{2}[\phi(x, y) + \phi(-x, -y)]$$

for all $x, y \in X$. From Theorem 3.1.5, it follows that there exists a unique cubic function $C : X \to Y$ satisfying Eq. [3.15]. Now it is obvious that Eq. [3.23] holds true for all $x \in X$ and so the proof of the theorem is complete. □

Corollary 3.1.7 (Chang and Jung [23]). *Let X be a real vector space and Y be a real Banach space. Let θ, p be real numbers such that $\theta \geq 0$ and $p < 2$. If a function $f : X \to Y$ satisfies the inequality*

$$\|Df(x, y)\| \leq \theta(\|x\|^p + \|y\|^p)$$

for all $x, y \in X$ and $f(0) = 0$, then there exists a unique cubic function $C : X \to Y$ and a unique quadratic function $Q : X \to Y$ satisfying Eq. [3.3] and

$$\|f(x) - C(x) - Q(x)\| \leq 3\left[\frac{1}{4 - 2^p} + \frac{1}{8 - 2^p}\right]\theta\|x\|^p$$

for all $x \in X$.

Corollary 3.1.8 (Chang and Jung [23]). *Let X be a real vector space and Y be a real Banach space. Let $\varepsilon \geq 0$ be a real number. If a function $f : X \to Y$ satisfies the inequality*

$$\|Df(x, y)\| \leq \varepsilon$$

for all $x, y \in X$, then there exist a unique cubic function $C : X \to Y$ and a unique quadratic function $Q : X \to Y$ satisfying Eq. [3.3] and

$$\|f(x) - C(x) - Q(x)\| \leq \frac{50}{63}\varepsilon$$

for all $x \in X$.

Jun and Kim [24] obtained the generalized Hyers-Ulam stability for a mixed type of cubic and additive functional equation. They considered the functional equation

$$
\begin{aligned}
f\left(\left(\sum_{i=1}^{l} x_i + x_{l+1}\right)\right) + f\left(\left(\sum_{i=1}^{l} x_i - x_{l+1}\right)\right) + 2\sum_{i=1}^{l} f(x_i) \\
= 2f\left(\sum_{i=1}^{l} x_i\right) + \sum_{i=1}^{l} [f(x_i + x_{l+1}) + f(x_i - x_{l+1})],
\end{aligned}
\tag{3.24}
$$

where l is a positive integer with $l \geq 2$. Let both E_1 and E_2 be real vector spaces. A mapping $f : E_1 \to E_2$ satisfies the functional equation [3.24] if and only if there exist two mappings $G : E_1 \times E_1 \times E_1 \to E_2$ and $A : E_1 \to E_2$ and a constant c in E_2 such that

$$f(x) = G(x, x, x) + A(x) + c$$

for all $x \in E_1$, where A is additive, and G is symmetric for each fixed one variable and additive for each fixed two variables. Additionally they solved the generalized Hyers-Ulam stability problem for Eq. [3.24] under the approximately cubic (or additive) condition by the iterative methods.

The generalized Hyers-Ulam stability for a mixed type of quadratic and additive functional equation in quasi-Banach spaces has been investigated by Najati and Moghimi [25]. They dealt with the functional equation

$$f(2x + y) + f(2x - y) = f(x + y) + f(x - y) + 2f(2x) - 2f(x). \tag{3.25}$$

It is easy to see that the function $f(x) = ax^2 + bx + c$ is a solution of the functional equation [3.25]. They investigated the generalized Hyers-Ulam stability for Eq. [3.25] in quasi-Banach spaces.

In [26], Prak and Bae considered the following quartic functional equation:

$$f(x + 2y) + f(x - 2y) = 4(f(x + y) + f(x - y) + 6f(y)) - 6f(x). \tag{3.26}$$

In fact, they proved that a function f between two real vector spaces X and Y is a solution of Eq. [3.26] if and only if there exists a unique symmetric multi-additive function $D : X \times X \times X \times X \to Y$ such that $f(x) = D(x, x, x, x)$ for all $x \in X$. It is easy to show that the function $f(x) = x^4$ satisfies the functional equation [3.26], which is called a quartic functional equation.

Eshaghi et al. [27] established the general solution of the functional equation

$$f(nx+y)+f(nx-y) = n^2f(x+y)+n^2f(x-y)+2(f(nx)-n^2f(x))-2(n^2-1)f(y) \tag{3.27}$$

for fixed integers n with $n \neq 0, \pm 1$ and investigate the generalized Hyers-Ulam stability of this equation in quasi-Banach spaces. It is easy to see that the function $f(x) = ax^4 + bx^2$ is a solution of the functional equation [3.27].

Lemma 3.1.9 (Eshaghi et al. [27]). *Let X and Y be real vector spaces. If a function* $f : X \to Y$ *satisfies the functional equation* [3.27], *then* f *is a quadratic and quartic function.*

Theorem 3.1.10 (Eshaghi et al. [27]). *A function* $f : X \to Y$ *satisfies Eq. [3.27] if and only if there exist a unique symmetric multi-additive function* $D : X \times X \times X \times X \to Y$ *and a unique symmetric bi-additive function* $B : X \times X \to Y$ *such that*

$$f(x) = D(x, x, x, x) + B(x, x)$$

for all $x \in X$.

Proof. We first assume that the function $f : X \to Y$ satisfies Eq. [3.27]. Let $g, h : X \to Y$ be functions defined by

$$g(x) := f(2x) - 16f(x), \quad h(x) := f(2x) - 4f(x)$$

for all $x \in X$. Hence, by Lemma 3.1.9, we achieve that the functions g and h are quadratic and quartic, respectively, and

$$f(x) := \frac{1}{12}h(x) - \frac{1}{12}g(x)$$

for all $x \in X$. Therefore, there exist a unique symmetric multi-additive mapping $D : X \times X \times X \times X \to Y$ and a unique symmetric bi-additive mapping $B : X \times X \to Y$ such that $D(x, x, x, x) = \frac{1}{12}h(x)$ and $B(x, x) = -\frac{1}{12}g(x)$ for all $x \in X$. So

$$f(x) = D(x, x, x, x) + B(x, x)$$

for all $x \in X$.

Conversely, assume that

$$f(x) = D(x, x, x, x) + B(x, x)$$

for all $x \in X$, where the function $D : X \times X \times X \times X \to Y$ is symmetric multi-additive and $B : X \times X \to Y$ is bi-additive defined above. By a simple computation, one can show that the functions D and B satisfy the functional equation [3.27], so the function f satisfies Eq. [3.27]. □

3.2 Ternary mixtures of functional equations

Eshaghi and Khodaei [28] introduce the following functional equation for fixed integers k with $k \neq 0, \pm 1$:

$$f(x + ky) + f(x - ky) = k^2 f(x + y) + k^2 f(x - y) + 2(1 - k^2) f(x), \tag{3.28}$$

with $f(0) = 0$. It is easy to see that the function $f(x) = ax^3 + bx^2 + cx$ is a solution of the functional equation [3.28]. In this section, we investigate the general solution of functional equation [3.28] when f is a mapping between vector spaces, and we establish the generalized Hyers-Ulam stability of the functional equation [3.28] whenever f is a function between two quasi-Banach spaces.

Lemma 3.2.1 (Eshaghi and Khodaei [28]). *Let X and Y be real vector spaces. If an even function $f : X \to Y$ with $f(0) = 0$ satisfies Eq. [3.28], then f is quadratic.*

Proof. Setting $x = 0$ in Eq. [3.28], by evenness of f, we obtain $f(kx) = k^2 f(x)$ for all $x \in X$. Replacing x by kx in Eq. [3.28] and then using the identity $f(kx) = k^2 f(x)$, we lead to

$$f(kx + y) + f(kx - y) = f(x + y) + f(x - y) + 2(k^2 - 1)f(x) \tag{3.29}$$

for all $x, y \in X$. Interchanging x with y in Eq. [3.28] gives

$$f(y + kx) + f(y - kx) = k^2 f(y + x) + k^2 f(y - x) + 2(1 - k^2)f(y) \tag{3.30}$$

for all $x, y \in X$. By evenness of f, it follows from Eq. [3.30] that

$$f(kx + y) + f(kx - y) = k^2 f(x + y) + k^2 f(x - y) + 2(1 - k^2)f(y) \tag{3.31}$$

for all $x, y \in X$. But, $k \neq 0, \pm 1$ so from Eqs. [3.29] and [3.31], we obtain

$$f(x + y) + f(x - y) = 2f(x) + 2f(y)$$

for all $x, y \in X$. This shows that f is quadratic, which completes the proof of lemma. □

Lemma 3.2.2 (Eshaghi and Khodaei [28]). *If an odd function $f : X \to Y$ satisfies Eq. [3.28], then f is cubic-additive.*

Proof. Letting $y = x$ in Eq. [3.28], we get, by oddness of f,

$$f((k + 1)x) = f((k - 1)x) + k^2 f(2x) + 2(1 - k^2)f(x) \tag{3.32}$$

for all $x, y \in X$. Replacing x by $(k - 1)x$ in Eq. [3.28] gives

$$\begin{aligned} &f((k - 1)x + ky) + f((k - 1)x - ky) \\ &\quad = k^2 f((k - 1)x + y) + k^2 f((k - 1)x - y) + 2(1 - k^2)f((k - 1)x) \end{aligned} \tag{3.33}$$

for all $x, y \in X$. Now, if we replace x with $(k + 1)x$ in Eq. [3.28] and use Eq. [3.32], we see that

$$\begin{aligned} &f((k + 1)x + ky) + f((k + 1)x - ky) \\ &\quad = k^2 f((k + 1)x + y) + k^2 f((k + 1)x - y) + 2(1 - k^2)f((k - 1)x) \\ &\qquad + 2k^2(1 - k^2)f(2x) + 4(1 - k^2)^2 f(x) \end{aligned} \tag{3.34}$$

for all $x, y \in X$. We substitute $x+y$ in place of x in Eq. [3.28] and then $x-y$ in place of x in Eq. [3.28] to obtain that

$$f(x+(k+1)y)+f(x-(k-1)y)=k^2f(x+2y)+2(1-k^2)f(x+y)+k^2f(x) \quad (3.35)$$

and

$$f(x-(k+1)y)+f(x+(k-1)y)=k^2f(x-2y)+2(1-k^2)f(x-y)+k^2f(x) \quad (3.36)$$

for all $x, y \in X$. If we subtract Eq. [3.36] from Eq. [3.35], we have

$$\begin{aligned} &f(x+(k+1)y)-f(x-(k+1)y) \\ &\quad = k^2f(x+2y)-k^2f(x-2y)+f(x+(k-1)y)-f(x-(k-1)y) \\ &\qquad +2(1-k^2)f(x+y)-2(1-k^2)f(x-y) \end{aligned} \quad (3.37)$$

for all $x, y \in X$. Interchanging x with y in Eq. [3.37] and using oddness of f, we get the relation

$$\begin{aligned} &f((k+1)x+y)+f((k+1)x-y) \\ &\quad = k^2f(2x+y)+k^2f(2x-y)+f((k-1)x+y)+f((k-1)x-y) \\ &\qquad +2(1-k^2)f(x+y)+2(1-k^2)f(x-y) \end{aligned} \quad (3.38)$$

for all $x, y \in X$. It follows from Eqs. [3.34] and [3.38] that

$$\begin{aligned} &f((k+1)x+ky)+f((k+1)x-ky) \\ &\quad = k^2f((k-1)x+y)+k^2f((k-1)x-y)+k^4f(2x+y)+k^4f(2x-y) \\ &\qquad +2k^2(1-k^2)f(x+y)+2k^2(1-k^2)f(x-y) \\ &\qquad +2(1-k^2)f((k-1)x)+2k^2(1-k^2)f(2x)+4(1-k^2)^2f(x) \end{aligned} \quad (3.39)$$

for all $x, y \in X$. We substitute $x+y$ in place of y in Eq. [3.28] and then $x-y$ in place of y in Eq. [3.28], and we get, by the oddness of f,

$$f((k+1)x+ky)-f((k-1)x+ky)=k^2f(2x+y)+k^2f(-y)+2(1-k^2)f(x) \quad (3.40)$$

and

$$f((k+1)x-ky)-f((k-1)x-ky)=k^2f(2x-y)+k^2f(y)+2(1-k^2)f(x) \quad (3.41)$$

for all $x, y \in X$. Then, by adding Eq. [3.40] to Eq. [3.41] and then using Eq. [3.33], we lead to

$$\begin{aligned} &f((k+1)x+ky)+f((k+1)x-ky) \\ &\quad = k^2f((k-1)x+y)+k^2f((k-1)x-y)+2(1-k^2)f((k-1)x) \\ &\qquad +k^2f(2x+y)+k^2f(2x-y)+4(1-k^2)f(x) \end{aligned} \quad (3.42)$$

for all $x, y \in X$. Finally, if we compare Eq. [3.39] with Eq. [3.42], then we conclude that

$$f(2x+y)+f(2x-y) = 2f(x+y)+2f(x-y)+2(f(2x)-2f(x))$$

for all $x, y \in X$. Hence, f is a cubic-additive function (see Section 3.1). This completes the proof. □

Theorem 3.2.3 (Eshaghi and Khodaei [28]). *A function $f : X \to Y$ with $f(0) = 0$ satisfies Eq. [3.28] for all $x, y \in X$ if and only if there exist functions $G : X \times X \times X \to Y$, $B : X \times X \to Y$ and $A : X \to Y$ such that $f(x) = G(x,x,x) + B(x,x) + A(x)$ for all $x \in X$, where the function G is symmetric for each fixed one variable and is additive for fixed two variables, and B is symmetric bi-additive and A is additive.*

Proof. Let f with $f(0) = 0$ satisfies Eq. [3.28]. We decompose f into the even part and odd part by putting

$$f_e(x) = \frac{1}{2}(f(x)+f(-x)), \quad f_o(x) = \frac{1}{2}(f(x)-f(-x)),$$

for all $x \in X$. It is clear that $f(x) = f_e(x) + f_o(x)$ for all $x \in X$. It is easy to show that the functions f_e and f_o satisfy Eq. [3.28]. Hence by Lemmas 3.2.1 and 3.2.2, we achieve that the functions f_e and f_o are quadratic and cubic-additive, respectively, thus there exist a symmetric bi-additive function $B : X \times X \to Y$ such that $f_e(x) = B(x,x)$ for all $x \in X$, and the function $G : X \times X \times X \to Y$ and additive function $A : X \to Y$ such that $f_o(x) = G(x,x,x)+A(x)$, for all $x \in X$, where the function G is symmetric for each fixed one variable and is additive for fixed two variables. Hence, we get $f(x) = G(x,x,x) + B(x,x) + A(x)$, for all $x \in X$.

Conversely, let $f(x) = G(x,x,x)+B(x,x)+A(x)$ for all $x \in X$, where the function G is symmetric for each fixed one variable and is additive for fixed two variables, and B is bi-additive and A is additive. By a simple computation one can show that the functions $x \mapsto G(x,x,x)$ and $x \mapsto B(x,x)$ and A satisfy the functional equation [3.28]. So the function f satisfies Eq. [3.28]. □

We now recall some basic facts concerning quasi-Banach space and some preliminary results.

Definition 3.2.4 ([18]). Let X be a real linear space. A quasi-norm is a real-valued function on X satisfying the following:

1. $\|x\| \geq 0$ for all $x \in X$ and $\|x\| = 0$ if and only if $x = 0$;
2. $\|\lambda \cdot x\| = |\lambda| \cdot \|x\|$ for all $\lambda \in \mathbb{R}$ and all $x \in X$;
3. There is a constant $M \geq 1$ such that $\|x + y\| \leq M(\|x\| + \|y\|)$ for all $x, y \in X$.

It follows from condition (3) that

$$\left\| \sum_{i=1}^{2n} x_i \right\| \leq M^n \sum_{i=1}^{2n} \|x_i\|, \quad \left\| \sum_{i=1}^{2n+1} x_i \right\| \leq M^{n+1} \sum_{i=1}^{2n+1} \|x_i\|$$

for all $n \geq 1$ and all $x_1, x_2, \ldots, x_{2n+1} \in X$.

The pair $(X, \|\cdot\|)$ is called a quasi-normed space if $\|\cdot\|$ is a quasi-norm on X. The smallest possible M is called the modulus of concavity of $\|\cdot\|$. A quasi-Banach space is a complete quasi-normed space.

A quasi-norm $\|\cdot\|$ is called a p-norm $(0 < p \le 1)$ if

$$\|x+y\|^p \le \|x\|^p + \|y\|^p$$

for all $x, y \in X$. In this case, a quasi-Banach space is called a p-Banach space.

Given a p-norm, the formula $d(x, y) := \|x - y\|^p$ gives us a translation invariant metric on X. By the Aoki-Rolewicz Theorem [29] (see also [30]), each quasi-norm is equivalent to some p-norm. Since it is much easier to work with p-norms, we henceforth restrict our attention mainly to p-norms. Moreover, J. Tabor [31] has investigated a version of Hyers-Rassias-Gajda theorem in quasi-Banach spaces.

Lemma 3.2.5 ([18]). *Let $0 < p \le 1$ and let $x_1, x_2, \ldots, x_n$ be nonnegative real numbers. Then*

$$\left(\sum_{i=1}^{n} x_i\right)^p \le \sum_{i=1}^{n} x_i^p.$$

For given $f : X \to Y$, we define the difference operator $D_f : X \times X \to Y$ by

$$D_f(x, y) = f(x+ky) + f(x-ky) - k^2 f(x+y) - k^2 f(x-y) - 2(1-k^2) f(x)$$

for all $x, y \in X$. Let $\varphi_-^p(x, y) := (\varphi_-(x, y))^p$.

Theorem 3.2.6 (Eshaghi and Khodaei [28]). *Let $j \in \{-1, 1\}$ be fixed and let $\varphi : X \times X \to [0, \infty)$ be a function such that*

$$\lim_{n\to\infty} k^{2nj} \varphi\left(\frac{x}{k^{nj}}, \frac{y}{k^{nj}}\right) = 0 \tag{3.43}$$

for all $x, y \in X$ and

$$\tilde{\psi}_e(x) := \sum_{i=\frac{1+j}{2}}^{\infty} k^{2ipj} \varphi^p\left(0, \frac{x}{k^{ij}}\right) < \infty \tag{3.44}$$

for all $x \in X$. Suppose that an even function $f : X \to Y$ with $f(0) = 0$ satisfies the inequality

$$\|D_f(x, y)\|_Y \le \varphi(x, y) \tag{3.45}$$

for all $x, y \in X$. Then the limit

$$Q(x) := \lim_{n\to\infty} k^{2nj} f\left(\frac{x}{k^{nj}}\right) \tag{3.46}$$

exists for all $x \in X$ and $Q : X \to Y$ is a unique quadratic function satisfying

$$\|f(x) - Q(x)\|_Y \leq \frac{M}{2k^2}[\tilde{\psi}_e(x)]^{1/p} \tag{3.47}$$

for all $x \in X$.

Proof. Let $j = 1$. By putting $x = 0$ in [3.45], we get

$$\|2f(ky) - 2k^2 f(y)\|_Y \leq \varphi(0, y) \tag{3.48}$$

for all $y \in X$. If we replace y in Eq. [3.48] by x, and divide both sides of Eq. [3.48] by 2, we get

$$\|f(kx) - k^2 f(x)\|_Y \leq \frac{1}{2}\varphi(0, x) \tag{3.49}$$

for all $x \in X$. Let $\psi_e(x) = \frac{1}{2}\varphi(0, x)$ for all $x \in X$, then by Eq. [3.49] we get

$$\|f(kx) - k^2 f(x)\|_Y \leq \psi_e(x) \tag{3.50}$$

for all $x \in X$. If we replace x in Eq. [3.50] by $\frac{x}{k^{n+1}}$ and multiply both sides of Eq. [3.50] by k^{2n}, then we have

$$\left\|k^{2(n+1)} f\left(\frac{x}{k^{n+1}}\right) - k^{2n} f\left(\frac{x}{k^n}\right)\right\|_Y \leq M k^{2n} \psi_e\left(\frac{x}{k^{n+1}}\right) \tag{3.51}$$

for all $x \in X$ and all nonnegative integers n. Since Y is p-Banach space, then Eq. [3.51] gives

$$\begin{aligned} \left\|k^{2(n+1)} f\left(\frac{x}{k^{n+1}}\right) - k^{2m} f\left(\frac{x}{k^m}\right)\right\|_Y^p &\leq \sum_{i=m}^{n} \left\|k^{2(i+1)} f\left(\frac{x}{k^{i+1}}\right) - k^{2i} f\left(\frac{x}{k^i}\right)\right\|_Y^p \\ &\leq M^p \sum_{i=m}^{n} k^{2ip} \psi_e{}^p\left(\frac{x}{k^{i+1}}\right) \end{aligned} \tag{3.52}$$

for all nonnegative integers n and m with $n \geq m$ and all $x \in X$. Since $\psi_e{}^p(x) = \frac{1}{2^p}\varphi^p(0, x)$ for all $x \in X$, therefore by Eq. [3.44] we have

$$\sum_{i=1}^{\infty} k^{2ip} \psi_e{}^p\left(\frac{x}{k^i}\right) < \infty \tag{3.53}$$

for all $x \in X$. Therefore we conclude from Eqs. [3.52] and [3.53] that the sequence $\{k^{2n} f\left(\frac{x}{k^n}\right)\}$ is a Cauchy sequence for all $x \in X$. Since Y is complete, the sequence $\{k^{2n} f\left(\frac{x}{k^n}\right)\}$ converges for all $x \in X$. So one can define the function $Q : X \to Y$ by Eq. [3.46] for all $x \in X$. Letting $m = 0$ and passing the limit $n \to \infty$ in Eq. [3.52], we get

$$\|f(x) - Q(x)\|_Y^p \le M^p \sum_{i=0}^{\infty} k^{2ip} \psi_e{}^p \left(\frac{x}{k^{i+1}}\right) = \frac{M^p}{k^{2p}} \sum_{i=1}^{\infty} k^{2ip} \psi_e{}^p \left(\frac{x}{k^i}\right) \tag{3.54}$$

for all $x \in X$. Therefore Eq. [3.47] follows from Eqs. [3.44] and [3.54]. Now we show that Q is quadratic. It follows from Eqs. [3.43], [3.45], and [3.46] that

$$\|D_Q(x,y)\|_Y = \lim_{n\to\infty} k^{2n} \|D_f\left(\frac{x}{k^n}, \frac{y}{k^n}\right)\|_Y \le \lim_{n\to\infty} k^{2n} \varphi\left(\frac{x}{k^n}, \frac{y}{k^n}\right) = 0$$

for all $x, y \in X$. Therefore the function $Q : X \to Y$ satisfies Eq. [3.28]. Since f is an even function, then Eq. [3.46] implies that the function $Q : X \to Y$ is even. Therefore by Lemma 3.2.1, we get that the function $Q : X \to Y$ is quadratic.

To prove the uniqueness of Q, let $Q' : X \to Y$ be another quadratic function satisfying Eq. [3.47]. Since

$$\lim_{n\to\infty} k^{2np} \sum_{i=1}^{\infty} k^{2ip} \varphi^p \left(0, \frac{x}{k^{i+n}}\right) = \lim_{n\to\infty} \sum_{i=n+1}^{\infty} k^{2ip} \varphi^p \left(0, \frac{x}{k^i}\right) = 0$$

for all $x \in X$, then

$$\lim_{n\to\infty} k^{2np} \tilde{\psi}_e \left(\frac{x}{k^n}\right) = 0$$

for all $x \in X$. Therefore it follows from Eq. [3.47] and the last equation that

$$\begin{aligned}\|Q(x) - Q'(x)\|_Y^p &= \lim_{n\to\infty} k^{2np} \left\| f\left(\frac{x}{k^n}\right) - Q'\left(\frac{x}{k^n}\right) \right\|_Y^p \\ &\le \frac{M^p}{2k^{2p}} \lim_{n\to\infty} k^{2np} \tilde{\psi}_e \left(\frac{x}{k^n}\right) = 0\end{aligned}$$

for all $x \in X$. Hence $Q = Q'$.

For $j = -1$, we can prove the theorem by a similar technique. □

Corollary 3.2.7 (Eshaghi and Khodaei [28]). *Let θ, r, s be nonnegative real numbers such that $r, s > 2$ or $0 \le r, s < 2$. Suppose that an even function $f : X \to Y$ with $f(0) = 0$ satisfies the inequality*

$$\|D_f(x,y)\|_Y \le \theta(\|x\|_X^r + \|y\|_X^s) \tag{3.55}$$

for all $x, y \in X$. Then there exists a unique quadratic function $Q : X \to Y$ satisfying

$$\|f(x) - Q(x)\|_Y \le \frac{M\theta}{2} \left(\frac{1}{|k^{2p} - k^{sp}|} \|x\|_X^{sp}\right)^{1/p}$$

for all $x \in X$.

Proof. It follows from Theorem 3.2.6 by putting $\varphi(x,y) := \theta(\|x\|_X^r + \|y\|_X^s)$ for all $x, y \in X$. □

Theorem 3.2.8 (Eshaghi and Khodaei [28]). *Let $j \in \{-1, 1\}$ be fixed and let $\varphi_a : X \times X \to [0, \infty)$ be a function such that*

$$\lim_{n\to\infty} 2^{nj}\varphi_a\left(\frac{x}{2^{nj}}, \frac{y}{2^{nj}}\right) = 0 \tag{3.56}$$

for all $x, y \in X$ and

$$\sum_{i=\frac{1+j}{2}}^{\infty} 2^{ipj}\varphi_a^p\left(\frac{u}{2^{ij}}, \frac{y}{2^{ij}}\right) < \infty \tag{3.57}$$

for all $u \in \{x, 2x, (k-1)x, (k+1)x, (2k-1)x, (2k+1)x\}$ and all $y \in \{x, 2x, 3x\}$. Suppose that an odd function $f : X \to Y$ satisfies the inequality

$$\|D_f(x, y)\|_Y \le \varphi_a(x, y) \tag{3.58}$$

for all $x, y \in X$. Then the limit

$$A(x) := \lim_{n\to\infty} 2^{nj}\left[f\left(\frac{x}{2^{nj-1}}\right) - 8f\left(\frac{x}{2^{nj}}\right)\right] \tag{3.59}$$

exists for all $x \in X$ and $A : X \to Y$ is a unique additive function satisfying

$$\|f(2x) - 8f(x) - A(x)\|_Y \le \frac{M^4}{2}[\tilde{\psi}_a(x)]^{1/p} \tag{3.60}$$

for all $x \in X$, where

$$\begin{aligned}\tilde{\psi}_a(x) := \sum_{i=\frac{1+j}{2}}^{\infty} 2^{ipj} \Bigg\{ &\frac{1}{k^{2p}(k^2-1)^p}\Bigg[(4k^2-3)^p\varphi_a^p\left(\frac{x}{2^{ij}}, \frac{x}{2^{ij}}\right) \\ &+ k^{2p}\varphi_a^p\left(\frac{2x}{2^{ij}}, \frac{2x}{2^{ij}}\right) + (2k^2)^p\varphi_a^p\left(\frac{2x}{2^{ij}}, \frac{x}{2^{ij}}\right) + (2k^2)^p\varphi_a^p\left(\frac{x}{2^{ij}}, \frac{2x}{2^{ij}}\right) \\ &+ \varphi_a^p\left(\frac{x}{2^{ij}}, \frac{3x}{2^{ij}}\right) + 2^p\varphi_a^p\left(\frac{(k+1)x}{2^{ij}}, \frac{x}{2^{ij}}\right) + 2^p\varphi_a^p\left(\frac{(k-1)x}{2^{ij}}, \frac{x}{2^{ij}}\right) \\ &+\varphi_a^p\left(\frac{(2k+1)x}{2^{ij}}, \frac{x}{2^{ij}}\right) + \varphi_a^p\left(\frac{(2k-1)x}{2^{ij}}, \frac{x}{2^{ij}}\right)\Bigg]\Bigg\}.\end{aligned} \tag{3.61}$$

Proof. Let $j = 1$. It follows from Eq. [3.58] and using oddness of f that

$$\|f(ky+x)-f(ky-x)-k^2f(x+y)-k^2f(x-y)+2(k^2-1)f(x)\| \le \varphi_a(x, y) \tag{3.62}$$

for all $x, y \in X$. Putting $y = x$ in Eq. [3.62], we have

$$\|f((k+1)x) - f((k-1)x) - k^2 f(2x) + 2(k^2-1)f(x)\| \leq \varphi_a(x, x) \tag{3.63}$$

for all $x \in X$. It follows from Eq. [3.63] that

$$\|f(2(k+1)x) - f(2(k-1)x) - k^2 f(4x) + 2(k^2-1)f(2x)\| \leq \varphi_a(2x, 2x) \tag{3.64}$$

for all $x \in X$. Replacing x and y by $2x$ and x in Eq. [3.62], respectively, we get

$$\|f((k+2)x) - f((k-2)x) - k^2 f(3x) - k^2 f(x) + 2(k^2-1)f(2x)\| \leq \varphi_a(2x, x) \tag{3.65}$$

for all $x \in X$. Setting $y = 2x$ in Eq. [3.62] gives

$$\|f((2k+1)x) - f((2k-1)x) - k^2 f(3x) - k^2 f(-x) + 2(k^2-1)f(x)\| \leq \varphi_a(x, 2x) \tag{3.66}$$

for all $x \in X$. Putting $y = 3x$ in Eq. [3.62], we obtain

$$\begin{aligned} &\|f((3k+1)x) - f((3k-1)x) - k^2 f(4x) - k^2 f(-2x) + 2(k^2-1)f(x)\| \\ &\quad \leq \varphi_a(x, 3x) \end{aligned} \tag{3.67}$$

for all $x \in X$. Replacing x and y by $(k+1)x$ and x in Eq. [3.62], respectively, we get

$$\begin{aligned} &\|f((2k+1)x) - f(-x) - k^2 f((k+2)x) - k^2 f(kx) + 2(k^2-1)f((k+1)x)\| \\ &\quad \leq \varphi_a((k+1)x, x) \end{aligned} \tag{3.68}$$

for all $x \in X$. Replacing x and y by $(k-1)x$ and x in Eq. [3.62], respectively, one gets

$$\begin{aligned} &\|f((2k-1)x) - f(x) - k^2 f((k-2)x) - k^2 f(kx) + 2(k^2-1)f((k-1)x)\| \\ &\quad \leq \varphi_a((k-1)x, x) \end{aligned} \tag{3.69}$$

for all $x \in X$. Replacing x and y by $(2k+1)x$ and x in Eq. [3.62], respectively, we obtain

$$\begin{aligned} &\|f((3k+1)x) - f(-(k+1)x) - k^2 f(2(k+1)x) - k^2 f(2kx) + 2(k^2-1)f((2k+1)x)\| \\ &\quad \leq \varphi_a((2k+1)x, x) \end{aligned} \tag{3.70}$$

for all $x \in X$. Replacing x and y by $(2k-1)x$ and x in Eq. [3.62], respectively, we have

$$\begin{aligned} &\|f((3k-1)x) - f(-(k-1)x) - k^2 f(2(k-1)x) - k^2 f(2kx) + 2(k^2-1)f((2k-1)x)\| \\ &\quad \leq \varphi_a((2k-1)x, x) \end{aligned} \tag{3.71}$$

for all $x \in X$. It follows from Eqs. [3.63], [3.65], [3.66], [3.68], and [3.69] that

$$\|f(3x) - 4f(2x) + 5f(x)\| \le \frac{M^3}{k^2(k^2-1)} [2(k^2-1)\varphi_a(x,x) + k^2\varphi_a(2x,x) + \varphi_a(x,2x) + \varphi_a((k+1)x,x) + \varphi_a((k-1)x,x)] \tag{3.72}$$

for all $x \in X$. And, from Eqs. [3.63], [3.64], [3.66], [3.67], [3.70], and [3.71], we conclude that

$$\begin{aligned}&\|f(4x) - 2f(3x) - 2f(2x) + 6f(x)\| \\ &\le \frac{M^3}{k^2(k^2-1)}[\varphi_a(x,x) + k^2\varphi_a(2x,2x) + 2(k^2-1)\varphi_a(x,2x) \\ &+ \varphi_a(x,3x) + \varphi_a((2k+1)x,x) + \varphi_a((2k-1)x,x)]\end{aligned} \tag{3.73}$$

for all $x \in X$. Finally, by using Eqs. [3.72] and [3.73], we obtain

$$\begin{aligned}&\|f(4x) - 10f(2x) + 16f(x)\| \\ &\le \frac{M^4}{k^2(k^2-1)}[(4k^2-3)\varphi_a(x,x) + k^2\varphi_a(2x,2x) \\ &+ 2k^2\varphi_a(2x,x) + 2k^2\varphi_a(x,2x) + \varphi_a(x,3x) + 2\varphi_a((k+1)x,x) \\ &+ 2\varphi_a((k-1)x,x) + \varphi_a((2k+1)x,x) + \varphi_a((2k-1)x,x)]\end{aligned} \tag{3.74}$$

for all $x \in X$, and let

$$\begin{aligned}\psi_a(x) = &\frac{1}{k^2(k^2-1)}[(4k^2-3)\varphi_a(x,x) + k^2\varphi_a(2x,2x) \\ &+ 2k^2\varphi_a(2x,x) + 2k^2\varphi_a(x,2x) + \varphi_a(x,3x) + 2\varphi_a((k+1)x,x) \\ &+ 2\varphi_a((k-1)x,x) + \varphi_a((2k+1)x,x) + \varphi_a((2k-1)x,x)]\end{aligned} \tag{3.75}$$

for all $x \in X$. Thus Eq. [3.74] means that

$$\|f(4x) - 10f(2x) + 16f(x)\| \le M^4\psi_a(x) \tag{3.76}$$

for all $x \in X$. Let $g : X \to Y$ be a function defined by $g(x) := f(2x) - 8f(x)$ for all $x \in X$. From Eq. [3.76], we conclude that

$$\|g(2x) - 2g(x)\| \le M^4\psi_a(x) \tag{3.77}$$

for all $x \in X$. If we replace x in Eq. [3.77] by $\frac{x}{2^{n+1}}$ and multiply both sides of Eq. [3.77] by 2^n, we see that

$$\left\|2^{n+1}g\left(\frac{x}{2^{n+1}}\right) - 2^n g\left(\frac{x}{2^n}\right)\right\|_Y \le M^4 2^n \psi_a\left(\frac{x}{2^{n+1}}\right) \tag{3.78}$$

for all $x \in X$ and all nonnegative integers n. Since Y is a p-Banach space, the inequality [3.78] gives

$$\left\|2^{n+1} g\left(\frac{x}{2^{n+1}}\right)-2^{m} g\left(\frac{x}{2^{m}}\right)\right\|_{Y}^{p} \leq \sum_{i=m}^{n}\left\|2^{i+1} g\left(\frac{x}{2^{i+1}}\right)-2^{i} g\left(\frac{x}{2^{i}}\right)\right\|_{Y}^{p} \leq M^{4p} \sum_{i=m}^{n} 2^{ip} \psi_{a}^{p}\left(\frac{x}{2^{i+1}}\right) \tag{3.79}$$

for all nonnegative integers n and m with $n \geq m$ and all $x \in X$. Since $0 < p \leq 1$, so by Lemma 3.2.5 and Eq. [3.75] we get

$$\begin{aligned}\psi_{a}^{p}(x) \leq & \frac{1}{k^{2p}(k^{2}-1)^{p}}[(4k^{2}-3)^{p} \varphi_{a}^{p}(x, x)+k^{2p} \varphi_{a}^{p}(2x, 2x) \\ & +(2k^{2})^{p} \varphi_{a}^{p}(2x, x)+(2k^{2})^{p} \varphi_{a}^{p}(x, 2x)+\varphi_{a}^{p}(x, 3x)+2^{p} \varphi_{a}^{p}((k+1)x, x) \\ & +2^{p} \varphi_{a}^{p}((k-1)x, x)+\varphi_{a}^{p}((2k+1)x, x)+\varphi_{a}^{p}((2k-1)x, x)]\end{aligned} \tag{3.80}$$

for all $x \in X$. Therefore it follows from Eqs. [3.56], [3.57], and [3.80] that

$$\sum_{i=1}^{\infty} 2^{ip} \psi_{a}^{p}\left(\frac{x}{2^{i}}\right)<\infty, \quad \lim_{n \rightarrow \infty} 2^{n} \psi_{a}\left(\frac{x}{2^{n}}\right)=0 \tag{3.81}$$

for all $x \in X$. It follows from Eqs. [3.79] and [3.81] that the sequence $\{2^{n} g(\frac{x}{2^{n}})\}$ is Cauchy for all $x \in X$. Since Y is complete, the sequence $\{2^{n} g(\frac{x}{2^{n}})\}$ converges for all $x \in X$. So one can define a function $A : X \rightarrow Y$ by

$$A(x)=\lim_{n \rightarrow \infty} 2^{n} g\left(\frac{x}{2^{n}}\right) \tag{3.82}$$

for all $x \in X$. Letting $m = 0$ and passing the limit $n \rightarrow \infty$ in Eq. [3.79], we get

$$\|g(x)-A(x)\|_{Y}^{p} \leq M^{4p} \sum_{i=0}^{\infty} 2^{ip} \psi_{a}^{p}\left(\frac{x}{2^{i+1}}\right)=\frac{M^{4p}}{2^{p}} \sum_{i=1}^{\infty} 2^{ip} \psi_{a}^{p}\left(\frac{x}{2^{i}}\right) \tag{3.83}$$

for all $x \in X$. Therefore, Eq. [3.60] follows from Eqs. [3.80] and [3.83]. Now we show that A is additive. It follows from Eqs. [3.78], [3.81], and [3.82] that

$$\begin{aligned}\|A(2x)-2A(x)\|_{Y} &=\lim_{n \rightarrow \infty}\left\|2^{n} g\left(\frac{x}{2^{n-1}}\right)-2^{n+1} g\left(\frac{x}{2^{n}}\right)\right\|_{Y} \\ &=2 \lim_{n \rightarrow \infty}\left\|2^{n-1} g\left(\frac{x}{2^{n-1}}\right)-2^{n} g\left(\frac{x}{2^{n}}\right)\right\|_{Y} \\ & \leq M^{4} \lim_{n \rightarrow \infty} 2^{n} \psi_{a}\left(\frac{x}{2^{n}}\right)=0\end{aligned}$$

for all $x \in X$. So

$$A(2x) = 2A(x) \tag{3.84}$$

for all $x \in X$. On the other hand, it follows from Eqs. [3.56], [3.58], and [3.82] that

$$\begin{aligned}
\|D_A(x,y)\|_Y &= \lim_{n\to\infty} 2^n \left\| D_g\left(\frac{x}{2^n}, \frac{y}{2^n}\right)\right\|_Y \\
&= \lim_{n\to\infty} 2^n \left\| D_f\left(\frac{x}{2^{n-1}}, \frac{y}{2^{n-1}}\right) - 8D_f\left(\frac{x}{2^n}, \frac{y}{2^n}\right)\right\|_Y \\
&\leq M^4 \lim_{n\to\infty} 2^n \left\{ \left\| D_f\left(\frac{x}{2^{n-1}}, \frac{y}{2^{n-1}}\right)\right\|_Y + 8\left\| D_f\left(\frac{x}{2^n}, \frac{y}{2^n}\right)\right\|_Y \right\} \\
&\leq M^4 \lim_{n\to\infty} 2^n \left\{ \varphi_a\left(\frac{x}{2^{n-1}}, \frac{y}{2^{n-1}}\right) + 8\varphi_a\left(\frac{x}{2^n}, \frac{y}{2^n}\right)\right\} = 0
\end{aligned}$$

for all $x, y \in X$. Hence the function A satisfies Eq. [3.28]. Thus by Lemma 3.2.2, the function $x \rightsquigarrow A(2x) - 8A(x)$ is cubic-additive. Therefore, Eq. [3.84] implies that the function A is additive.

To prove the uniqueness property of A, let $A' : X \to Y$ be another additive function satisfying Eq. [3.60]. Since

$$\lim_{n\to\infty} 2^{np} \sum_{i=1}^{\infty} 2^{ip} \varphi_a^p \left(\frac{u}{2^{n+i}}, \frac{y}{2^{n+i}}\right) = \lim_{n\to\infty} \sum_{i=n+1}^{\infty} 2^{ip} \varphi_a^p \left(\frac{u}{2^i}, \frac{y}{2^i}\right) = 0$$

for all $u \in \{x, 2x, (k-1)x, (k+1)x, (2k-1)x, (2k+1)x\}$ and all $y \in \{x, 2x, 3x\}$. Hence

$$\lim_{n\to\infty} 2^{np} \tilde{\psi}_a\left(\frac{x}{2^n}\right) = 0 \tag{3.85}$$

for all $x \in X$. It follows from Eqs. [3.60] and [3.85] that

$$\|A(x) - A'(x)\|_Y^p = \lim_{n\to\infty} 2^{np} \left\| g\left(\frac{x}{2^n}\right) - A'\left(\frac{x}{2^n}\right)\right\|_Y^p \leq \frac{M^{4p}}{2^p} \lim_{n\to\infty} 2^{np} \tilde{\psi}_a\left(\frac{x}{2^n}\right) = 0$$

for all $x \in X$. So $A = A'$.

For $j = -1$, we can prove the theorem by a similar technique. □

Corollary 3.2.9 (Eshaghi and Khodaei [28]). *Let θ, r, s be nonnegative real numbers such that $r, s > 1$ or $r, s < 1$. Suppose that an odd function $f : X \to Y$ satisfies the inequality*

$$\|D_f(x,y)\|_Y \leq \begin{cases} \theta, & r = s = 0; \\ \theta \|x\|_X^r, & r > 0, s = 0; \\ \theta \|y\|_X^s, & r = 0, s > 0; \\ \theta(\|x\|_X^r + \|y\|_X^s), & r, s > 0, \end{cases} \tag{3.86}$$

for all $x, y \in X$. Then there exists a unique additive function $A : X \to Y$ satisfying

$$\|f(2x)-8f(x)-A(x)\|_Y \leq \frac{M^4\theta}{k^2(k^2-1)} \begin{cases} \delta_a, & r=s=0; \\ \alpha_a \|x\|_X^r, & r>0, s=0; \\ \beta_a \|x\|_X^s, & r=0, s>0; \\ (\alpha_a^p \|x\|_X^{rp} + \beta_a^p \|x\|_X^{sp})^{1/p}, & r,s>0, \end{cases}$$

for all $x \in X$, where

$$\delta_a = \left\{\frac{1}{2^p-1}\left[(4k^2-3)^p + 2^{p+1}(k^{2p}+1) + k^{2p} + 3\right]\right\}^{1/p},$$

$$\alpha_a = \left\{\frac{1}{|2^p-2^{rp}|}\left[(4k^2-3)^p + (2k+1)^{rp} + (2k-1)^{rp} + 2^p(k+1)^{rp}\right.\right.$$
$$\left.\left. +2^p(k-1)^{rp} + k^{2p}(2^{(r+1)p} + 2^{rp} + 2^p) + 1\right]\right\}^{1/p},$$

$$\beta_a = \left\{\frac{1}{|2^p-2^{sp}|}\left[(4k^2-3)^p + k^{2p}(2^{(s+1)p} + 2^{sp} + 2^p) + 3^{sp} + 2^{p+1} + 2\right]\right\}^{1/p}.$$

Corollary 3.2.10 (Eshaghi and Khodaei [28]). *Let $\theta \geq 0$ and $r, s > 0$ be nonnegative real numbers such that $\lambda := r+s \neq 1$. Suppose that an odd function $f : X \to Y$ satisfies the inequality*

$$\|D_f(x,y)\|_Y \leq \theta \|x\|_X^r \|y\|_X^s \tag{3.87}$$

for all $x, y \in X$. Then there exists a unique additive function $A : X \to Y$ satisfying

$$\|f(2x) - 8f(x) - A(x)\|_Y \leq \frac{M^4\theta}{k^2(k^2-1)} \varepsilon_a \|x\|_X^\lambda$$

for all $x \in X$, where

$$\varepsilon_a = \left\{\frac{1}{|2^p-2^{\lambda p}|}\left[(4k^2-3)^p + (2k+1)^{rp} + (2k-1)^{rp} + 2^p(k+1)^{rp}\right.\right.$$
$$\left.\left. +2^p(k-1)^{rp} + k^{2p}(2^{(r+1)p} + 2^{(s+1)p} + 2^{\lambda p}) + 3^{sp}\right]\right\}^{1/p}.$$

Theorem 3.2.11 (Eshaghi and Khodaei [28]). *Let $j \in \{-1, 1\}$ be fixed and let $\varphi_c : X \times X \to [0, \infty)$ be a function such that*

$$\lim_{n\to\infty} 8^{nj}\varphi_c\left(\frac{x}{2^{nj}}, \frac{y}{2^{nj}}\right) = 0 \tag{3.88}$$

for all $x, y \in X$ *and*

$$\sum_{i=\frac{1+j}{2}}^{\infty} 8^{ipj} \varphi_c^p \left(\frac{u}{2^{ij}}, \frac{y}{2^{ij}} \right) < \infty \tag{3.89}$$

for all $u \in \{x, 2x, (k-1)x, (k+1)x, (2k-1)x, (2k+1)x\}$ *and all* $y \in \{x, 2x, 3x\}$. *Suppose that an odd function* $f : X \to Y$ *satisfies the inequality*

$$\|D_f(x, y)\|_Y \leq \varphi_c(x, y) \tag{3.90}$$

for all $x, y \in X$. *Then the limit*

$$C(x) := \lim_{n\to\infty} 8^{nj} \left[f\left(\frac{x}{2^{nj-1}} \right) - 2f\left(\frac{x}{2^{nj}} \right) \right] \tag{3.91}$$

exists for all $x \in X$ *and* $C : X \to Y$ *is a unique cubic function satisfying*

$$\|f(2x) - 2f(x) - C(x)\|_Y \leq \frac{M^4}{8} [\tilde{\psi}_c(x)]^{1/p} \tag{3.92}$$

for all $x \in X$, *where*

$$\begin{aligned} \tilde{\psi}_c(x) := \sum_{i=\frac{1+j}{2}}^{\infty} 8^{ipj} \Bigg\{ & \frac{1}{k^{2p}(k^2-1)^p} \left[(4k^2-3)^p \varphi_c^p \left(\frac{x}{2^{ij}}, \frac{x}{2^{ij}} \right) \right. \\ & + k^{2p} \varphi_c^p \left(\frac{2x}{2^{ij}}, \frac{2x}{2^{ij}} \right) + (2k^2)^p \varphi_c^p \left(\frac{2x}{2^{ij}}, \frac{x}{2^{ij}} \right) + (2k^2)^p \varphi_c^p \left(\frac{x}{2^{ij}}, \frac{2x}{2^{ij}} \right) \\ & + \varphi_c^p \left(\frac{x}{2^{ij}}, \frac{3x}{2^{ij}} \right) + 2^p \varphi_c^p \left(\frac{(k+1)x}{2^{ij}}, \frac{x}{2^{ij}} \right) + 2^p \varphi_c^p \left(\frac{(k-1)x}{2^{ij}}, \frac{x}{2^{ij}} \right) \\ & \left. + \varphi_c^p \left(\frac{(2k+1)x}{2^{ij}}, \frac{x}{2^{ij}} \right) + \varphi_c^p \left(\frac{(2k-1)x}{2^{ij}}, \frac{x}{2^{ij}} \right) \right] \Bigg\}. \end{aligned} \tag{3.93}$$

Proof. Let $j = 1$. Similar to the proof of Theorem 3.2.8, we have

$$\|f(4x) - 10f(2x) + 16f(x)\| \leq M^4 \psi_c(x) \tag{3.94}$$

for all $x \in X$, where

$$\begin{aligned} \psi_c(x) = & \frac{1}{k^2(k^2-1)} [(4k^2-3)\varphi_c(x, x) + k^2 \varphi_c(2x, 2x) \\ & + 2k^2 \varphi_c(2x, x) + 2k^2 \varphi_c(x, 2x) + \varphi_c(x, 3x) + 2\varphi_c((k+1)x, x) \\ & + 2\varphi_c((k-1)x, x) + \varphi_c((2k+1)x, x) + \varphi_c((2k-1)x, x)] \end{aligned} \tag{3.95}$$

for all $x \in X$. Let $h : X \to Y$ be a function defined by $h(x) := f(2x) - 2f(x)$. Then, we conclude that

$$\|h(2x) - 8h(x)\| \leq M^4 \psi_c(x) \tag{3.96}$$

for all $x \in X$. If we replace x in Eq. [3.96] $\frac{x}{2^{n+1}}$ and multiply both sides of Eq. [3.96] by 8^n, we get

$$\left\| 8^{n+1} h\left(\frac{x}{2^{n+1}}\right) - 8^n h\left(\frac{x}{2^n}\right) \right\|_Y \leq M^4 8^n \psi_c\left(\frac{x}{2^{n+1}}\right) \tag{3.97}$$

for all $x \in X$ and all nonnegative integers n. Since Y is p-Banach space, then by Eq. [3.97], we have

$$\begin{aligned} \left\| 8^{n+1} h\left(\frac{x}{2^{n+1}}\right) - 8^m h\left(\frac{x}{2^m}\right) \right\|_Y^p &\leq \sum_{i=m}^{n} \left\| 8^{i+1} h\left(\frac{x}{2^{i+1}}\right) - 8^i h\left(\frac{x}{2^i}\right) \right\|_Y^p \\ &\leq M^{4p} \sum_{i=m}^{n} 8^{ip} \psi_c{}^p\left(\frac{x}{2^{i+1}}\right) \end{aligned} \tag{3.98}$$

for all nonnegative integers n and m with $n \geq m$ and all $x \in X$. Since $0 < p \leq 1$, so by Lemma 3.2.5 and Eq. [3.95], we get

$$\begin{aligned} \psi_c^p(x) \leq \frac{1}{k^{2p}(k^2-1)^p}&[(4k^2-3)^p \varphi_c^p(x,x) + k^{2p}\varphi_c^p(2x,2x) \\ &+ (2k^2)^p \varphi_c^p(2x,x) + (2k^2)^p \varphi_c^p(x,2x) + \varphi_c^p(x,3x) + 2^p \varphi_c^p((k+1)x,x) \\ &+ 2^p \varphi_c^p((k-1)x,x) + \varphi_c^p((2k+1)x,x) + \varphi_c^p((2k-1)x,x)] \end{aligned} \tag{3.99}$$

for all $x \in X$. Therefore it follows from Eqs. [3.88], [3.89], and [3.99] that

$$\sum_{i=1}^{\infty} 2^{ip} \psi_c^p\left(\frac{x}{2^i}\right) < \infty, \quad \lim_{n\to\infty} 2^n \psi_c\left(\frac{x}{2^n}\right) = 0 \tag{3.100}$$

for all $x \in X$. Therefore, we conclude from Eqs. [3.98] and [3.100] that the sequence $\{8^n h(\frac{x}{2^n})\}$ is a Cauchy sequence for all $x \in X$. Since Y is complete, the sequence $\{8^n h(\frac{x}{2^n})\}$ converges for all $x \in X$. So one can define the function $C : X \to Y$ by

$$C(x) = \lim_{n\to\infty} 8^n h\left(\frac{x}{2^n}\right) \tag{3.101}$$

for all $x \in X$. Letting $m = 0$ and passing the limit $n \to \infty$ in Eq. [3.98], we get

$$\|h(x) - C(x)\|_Y^p \leq M^{4p} \sum_{i=0}^{\infty} 8^{ip} \psi_c{}^p\left(\frac{x}{2^{i+1}}\right) = \frac{M^{4p}}{8^p} \sum_{i=1}^{\infty} 8^{ip} \psi_c{}^p\left(\frac{x}{2^i}\right) \tag{3.102}$$

for all $x \in X$. Therefore, Eq. [3.92] follows from Eqs. [3.99] and [3.102]. Now we show that C is cubic. It follows from Eqs. [3.97], [3.100], and [3.101] that

$$\begin{aligned}\|C(2x) - 8C(x)\|_Y &= \lim_{n\to\infty} \left\|8^n h\left(\frac{x}{2^{n-1}}\right) - 8^{n+1} h\left(\frac{x}{2^n}\right)\right\|_Y \\ &= 8 \lim_{n\to\infty} \left\|8^{n-1} h\left(\frac{x}{2^{n-1}}\right) - 8^n h\left(\frac{x}{2^n}\right)\right\|_Y \\ &\leq M^4 \lim_{n\to\infty} 8^n \psi_c\left(\frac{x}{2^n}\right) = 0\end{aligned}$$

for all $x \in X$. So

$$C(2x) = 8C(x) \tag{3.103}$$

for all $x \in X$. On the other hand it follows from Eqs. [3.88], [3.90], and [3.101] that

$$\begin{aligned}\|D_C(x,y)\|_Y &= \lim_{n\to\infty} 8^n \left\|D_h\left(\frac{x}{2^n}, \frac{y}{2^n}\right)\right\|_Y \\ &= \lim_{n\to\infty} 8^n \left\|D_f\left(\frac{x}{2^{n-1}}, \frac{y}{2^{n-1}}\right) - 2D_f\left(\frac{x}{2^n}, \frac{y}{2^n}\right)\right\|_Y \\ &\leq M^4 \lim_{n\to\infty} 8^n \left\{\left\|D_f\left(\frac{x}{2^{n-1}}, \frac{y}{2^{n-1}}\right)\right\|_Y + 2\left\|D_f\left(\frac{x}{2^n}, \frac{y}{2^n}\right)\right\|_Y\right\} \\ &\leq M^4 \lim_{n\to\infty} 8^n \left\{\varphi_C\left(\frac{x}{2^{n-1}}, \frac{y}{2^{n-1}}\right) + 2\varphi_C\left(\frac{x}{2^n}, \frac{y}{2^n}\right)\right\} = 0\end{aligned}$$

for all $x, y \in X$. Hence the function C satisfies Eq. [3.28]. By Lemma 3.2.2, the function $x \rightsquigarrow C(2x) - 8C(x)$ is cubic-additive. Hence, Eq. [3.103] implies that function C is cubic.

To prove the uniqueness of C, let $C' : X \to Y$ be another additive function satisfying Eq. [3.92]. Since

$$\lim_{n\to\infty} 8^{np} \sum_{i=1}^{\infty} 8^{ip} \varphi_c{}^p\left(\frac{u}{2^{n+i}}, \frac{y}{2^{n+i}}\right) = \lim_{n\to\infty} \sum_{i=n+1}^{\infty} 8^{ip} \varphi_c{}^p\left(\frac{u}{2^i}, \frac{y}{2^i}\right) = 0$$

for all $u \in \{x, 2x, (k-1)x, (k+1)x, (2k-1)x, (2k+1)x\}$ and all $y \in \{x, 2x, 3x\}$. Hence,

$$\lim_{n\to\infty} 8^{np} \tilde{\psi}_c\left(\frac{x}{2^n}\right) = 0 \tag{3.104}$$

for all $x \in X$. It follows from Eqs. [3.92] and [3.104] that

$$\|C(x) - C'(x)\|_Y = \lim_{n\to\infty} 8^{np} \left\|h\left(\frac{x}{2^n}\right) - C'\left(\frac{x}{2^n}\right)\right\|_Y^p \leq \frac{M^{4p}}{8^p} \lim_{n\to\infty} 8^{np} \tilde{\psi}_c\left(\frac{x}{2^n}\right) = 0$$

for all $x \in X$. So $C = C'$.

For $j = -1$, we can prove the theorem by a similar technique. □

Corollary 3.2.12 (Eshaghi and Khodaei [28]). *Let θ, r, s be nonnegative real numbers such that $r, s > 3$ or $r, s < 3$. Suppose that an odd function $f : X \to Y$ satisfies the inequality* [3.86] *for all $x, y \in X$. Then there exists a unique cubic function $C : X \to Y$ satisfying*

$$\|f(2x)-2f(x)-C(x)\|_Y \leq \frac{M^4\theta}{k^2(k^2-1)} \begin{cases} \delta_c, & r=s=0; \\ \alpha_c \ \|x\|_X^r, & r>0, s=0; \\ \beta_c \ \|x\|_X^s, & r=0, s>0; \\ (\alpha_c^p \ \|x\|_X^{rp} + \beta_c^p \ \|x\|_X^{sp})^{1/p}, & r,s>0, \end{cases}$$

for all $x \in X$, where

$$\delta_c = \left\{\frac{1}{8^p-1}\left[(4k^2-3)^p + 2^{p+1}(k^{2p}+1) + k^{2p} + 3\right]\right\}^{1/p},$$

$$\alpha_c = \left\{\frac{1}{|8^p-2^{rp}|}\left[(4k^2-3)^p + (2k+1)^{rp} + (2k-1)^{rp} + 2^p(k+1)^{rp}\right.\right.$$

$$\left.\left.+\,2^p(k-1)^{rp} + k^{2p}(2^{(r+1)p} + 2^{rp} + 2^p) + 1\right]\right\}^{1/p},$$

$$\beta_c = \left\{\frac{1}{|8^p-2^{sp}|}\left[(4k^2-3)^p + k^{2p}(2^{(s+1)p} + 2^{sp} + 2^p) + 3^{sp} + 2^{p+1} + 2\right]\right\}^{1/p}.$$

Corollary 3.2.13 (Eshaghi and Khodaei [28]). *Let $\theta \geq 0$ and $r, s > 0$ be nonnegative real numbers such that $\lambda := r + s \neq 3$. Suppose that an odd function $f : X \to Y$ satisfies the inequality* [3.87] *for all $x, y \in X$. Then there exists a unique cubic function $C : X \to Y$ satisfying*

$$\|f(2x)-2f(x)-C(x)\|_Y \leq \frac{M^4\theta}{k^2(k^2-1)} \ \varepsilon_c \ \|x\|_X^{\lambda},$$

for all $x \in X$, where

$$\varepsilon_c = \left\{\frac{1}{|8^p-2^{\lambda p}|}\left[(4k^2-3)^p + (2k+1)^{rp} + (2k-1)^{rp} + 2^p(k+1)^{rp}\right.\right.$$

$$\left.\left.+2^p(k-1)^{rp} + k^{2p}(2^{(r+1)p} + 2^{(s+1)p} + 2^{\lambda p}) + 3^{sp}\right]\right\}^{1/p}.$$

Theorem 3.2.14 (Eshaghi and Khodaei [28]). *Let $j \in \{-1, 1\}$ be fixed and let $\varphi : X \times X \to [0, \infty)$ be a function such that*

$$\lim_{n\to\infty}\left\{\left(\frac{1+j}{2}\right)2^{nj}\varphi\left(\frac{x}{2^{nj}}, \frac{y}{2^{nj}}\right) + \left(\frac{1-j}{2}\right)8^{nj}\varphi\left(\frac{x}{2^{nj}}, \frac{y}{2^{nj}}\right)\right\} = 0$$
$$\left(\lim_{n\to\infty}\left\{\left(\tfrac{1-j}{2}\right)2^{nj}\varphi\left(\tfrac{x}{2^{nj}}, \tfrac{y}{2^{nj}}\right) + \left(\tfrac{1+j}{2}\right)8^{nj}\varphi\left(\tfrac{x}{2^{nj}}, \tfrac{y}{2^{nj}}\right)\right\} = 0\right) \tag{3.105}$$

for all $x, y \in X$ and

$$\sum_{i=\frac{1+j}{2}}^{\infty} \left\{ \left(\frac{1+j}{2}\right) 2^{ipj} \varphi^p \left(\frac{u}{2^{ij}}, \frac{y}{2^{ij}}\right) + \left(\frac{1-j}{2}\right) 8^{ipj} \varphi^p \left(\frac{u}{2^{ij}}, \frac{y}{2^{ij}}\right) \right\} < \infty$$
$$\left(\sum_{i=\frac{1+j}{2}}^{\infty} \left\{ \left(\frac{1-j}{2}\right) 2^{ipj} \varphi^p \left(\frac{u}{2^{ij}}, \frac{y}{2^{ij}}\right) + \left(\frac{1+j}{2}\right) 8^{ipj} \varphi^p \left(\frac{u}{2^{ij}}, \frac{y}{2^{ij}}\right) \right\} < \infty\right) \tag{3.106}$$

for all $u \in \{x, 2x, (k-1)x, (k+1)x, (2k-1)x, (2k+1)x\}$ and all $y \in \{x, 2x, 3x\}$. Suppose that an odd function $f : X \to Y$ satisfies the inequality [3.45] *for all $x, y \in X$. Then there exist a unique cubic function $C : X \to Y$ and a unique additive function $A : X \to Y$ such that*

$$\|f(x) - C(x) - A(x)\|_Y \leq \frac{M^5}{48} \left(4 \left[\tilde{\psi}_a(x)\right]^{1/p} + \left[\tilde{\psi}_c(x)\right]^{1/p} \right) \tag{3.107}$$

for all $x \in X$, where

$$\begin{aligned}
\tilde{\psi}_a(x) := \sum_{i=\frac{1+j}{2}}^{\infty} 2^{ipj} \Bigg\{ \frac{1}{k^{2p}(k^2-1)^p} \Bigg[&(4k^2-3)^p \varphi^p \left(\frac{x}{2^{ij}}, \frac{x}{2^{ij}}\right) + k^{2p} \varphi^p \left(\frac{2x}{2^{ij}}, \frac{2x}{2^{ij}}\right) \\
&+ (2k^2)^p \varphi^p \left(\frac{2x}{2^{ij}}, \frac{x}{2^{ij}}\right) + (2k^2)^p \varphi^p \left(\frac{x}{2^{ij}}, \frac{2x}{2^{ij}}\right) + \varphi^p \left(\frac{x}{2^{ij}}, \frac{3x}{2^{ij}}\right) \\
&+ 2^p \varphi^p \left(\frac{(k+1)x}{2^{ij}}, \frac{x}{2^{ij}}\right) + 2^p \varphi^p \left(\frac{(k-1)x}{2^{ij}}, \frac{x}{2^{ij}}\right) \\
&+ \varphi^p \left(\frac{(2k+1)x}{2^{ij}}, \frac{x}{2^{ij}}\right) + \varphi^p \left(\frac{(2k-1)x}{2^{ij}}, \frac{x}{2^{ij}}\right) \Bigg] \Bigg\},
\end{aligned} \tag{3.108}$$

$$\begin{aligned}
\tilde{\psi}_c(x) := \sum_{i=\frac{1+j}{2}}^{\infty} 8^{ipj} \Bigg\{ \frac{1}{k^{2p}(k^2-1)^p} \Bigg[&(4k^2-3)^p \varphi^p \left(\frac{x}{2^{ij}}, \frac{x}{2^{ij}}\right) + k^{2p} \varphi^p \left(\frac{2x}{2^{ij}}, \frac{2x}{2^{ij}}\right) \\
&+ (2k^2)^p \varphi^p \left(\frac{2x}{2^{ij}}, \frac{x}{2^{ij}}\right) + (2k^2)^p \varphi^p \left(\frac{x}{2^{ij}}, \frac{2x}{2^{ij}}\right) + \varphi^p \left(\frac{x}{2^{ij}}, \frac{3x}{2^{ij}}\right) \\
&+ 2^p \varphi^p \left(\frac{(k+1)x}{2^{ij}}, \frac{x}{2^{ij}}\right) + 2^p \varphi^p \left(\frac{(k-1)x}{2^{ij}}, \frac{x}{2^{ij}}\right) \\
&+ \varphi^p \left(\frac{(2k+1)x}{2^{ij}}, \frac{x}{2^{ij}}\right) + \varphi^p \left(\frac{(2k-1)x}{2^{ij}}, \frac{x}{2^{ij}}\right) \Bigg] \Bigg\}.
\end{aligned} \tag{3.109}$$

Proof. Let $j = 1$. By Theorems 3.2.8 and 3.2.11, there exist an additive function $A_0 : X \to Y$ and a cubic function $C_0 : X \to Y$ such that

$$\|f(2x) - 8f(x) - A_0(x)\|_Y \leq \frac{M^4}{2} \left[\tilde{\psi}_a(x)\right]^{1/p}, \quad \|f(2x) - 2f(x) - C_0(x)\|_Y \leq \frac{M^4}{8} \left[\tilde{\psi}_c(x)\right]^{1/p}$$

for all $x \in X$. Therefore, it follows from the last inequality that

$$\left\| f(x) + \frac{1}{6}A_0(x) - \frac{1}{6}C_0(x) \right\|_Y \le \frac{M^5}{48} \left(4\left[\tilde{\psi}_a(x)\right]^{1/p} + \left[\tilde{\psi}_c(x)\right]^{1/p} \right)$$

for all $x \in X$. So we obtain Eq. [3.65] by letting $A(x) = -\frac{1}{6}A_0(x)$ and $C(x) = \frac{1}{6}C_0(x)$ for all $x \in X$. To prove the uniqueness property of A and C, let $A_1, C_1 : X \to Y$ be another additive and cubic function satisfying Eq. [3.107]. Let $A' = A - A_1$ and $C' = C - C_1$. So

$$\begin{aligned} \|A'(x) + C'(x)\|_Y &\le M\{\|f(x) - A(x) - C(x)\|_Y + \|f(x) - A_1(x) - C_1(x)\|_Y\} \\ &\le \frac{M^6}{24} \left(4\left[\tilde{\psi}_a(x)\right]^{1/p} + \left[\tilde{\psi}_c(x)\right]^{1/p} \right) \end{aligned} \tag{3.110}$$

for all $x \in x$. Since

$$\lim_{n\to\infty} 2^{np}\tilde{\psi}_a\left(\frac{x}{2^n}\right) = \lim_{n\to\infty} 8^{np}\tilde{\psi}_c\left(\frac{x}{2^n}\right) = 0$$

for all $x \in X$, so if we replace x in Eq. [3.110] by $\frac{x}{2^n}$ and multiply both sides of Eq. [3.110] by 8^n, we get

$$\lim_{n\to\infty} 8^n \left\| A'\left(\frac{x}{2^n}\right) + C'\left(\frac{x}{2^n}\right) \right\|_Y = 0$$

for all $x \in X$. Therefore, $C' = 0$. So it follows from Eq. [3.110] that

$$\|A'(x)\|_Y \le \frac{5M^6}{24} \left[\tilde{\psi}_a(x)\right]^{1/p}$$

for all $x \in X$. Therefore $A' = 0$.

For $j = -1$, we can prove the theorem by a similar technique. □

Corollary 3.2.15 (Eshaghi and Khodaei [28]). *Let θ, r, s be nonnegative real numbers such that $r, s > 3$ or $1 < r, s < 3$ or $r, s < 1$. Suppose that an odd function $f : X \to Y$ satisfies the inequality* [3.86] *for all $x, y \in X$. Then there exists a unique additive function $A : X \to Y$ and a unique cubic function $C : X \to Y$ such that*

$$\|f(x) - A(x) - C(x)\|_Y \le \frac{M^5\theta}{6k^2(k^2-1)} \begin{cases} \delta_a + \delta_c, & r = s = 0; \\ (\alpha_a + \alpha_c)\ \|x\|_X^r, & r > 0, s = 0; \\ (\beta_a + \beta_c)\ \|x\|_X^s, & r = 0, s > 0; \\ \gamma_a(x) + \gamma_c(x), & r, s > 0. \end{cases}$$

for all $x \in X$, where $\delta_a, \delta_c, \alpha_a, \alpha_c, \beta_a$, and β_c are defined as in Corollaries 3.2.9 and 3.2.12, and

$$\gamma_a(x) = \{\alpha_a^p \|x\|_X^{rp} + \beta_a^p \|x\|_X^{sp}\}^{1/p}, \quad \gamma_c(x) = \{\alpha_c^p \|x\|_X^{rp} + \beta_c^p \|x\|_X^{sp}\}^{1/p}$$

for all $x \in X$.

Corollary 3.2.16 (Eshaghi and Khodaei [28]). *Let $\theta \geq 0$ and $r, s > 0$ be nonnegative real numbers such that $\lambda := r + s \in (0,1) \cup (1,3) \cup (3,\infty)$. Suppose that an odd function $f : X \to Y$ satisfies the inequality* [3.87] *for all $x, y \in X$. Then there exists a unique additive function $A : X \to Y$ and a unique cubic function $C : X \to Y$ such that*

$$\|f(x) - A(x) - C(x)\|_Y \leq \frac{M^5\theta}{6k^2(k^2-1)} \quad (\varepsilon_a + \varepsilon_c)\, \|x\|_X^{\lambda},$$

for all $x \in X$, where ε_a and ε_c are defined as in Corollaries 3.2.10 and 3.2.13.

Now, we are ready to prove the main theorem concerning the stability problem for the Eq. [3.28].

Theorem 3.2.17 (Eshaghi and Khodaei [28]). *Let $\varphi : X \times X \to [0,\infty)$ be a function which satisfies Eq. [3.43] for all $x, y \in X$ and Eq. [3.44] for all $x \in X$, or satisfies Eq. [3.105] for all $x, y \in X$ and Eq. [3.106] for all $u \in \{x, 2x, (k-1)x, (k+1)x, (2k-1)x, (2k+1)x\}$ and all $y \in \{x, 2x, 3x\}$. Suppose that a function $f : X \to Y$ satisfies the inequality* [3.45] *for all $x, y \in X$. Furthermore, assume that $f(0) = 0$ in Eq. [3.45] for the case f is even. Then there exist a unique cubic function $C : X \to Y$, a unique quadratic function $Q : X \to Y$ and a unique additive function $A : X \to Y$ such that*

$$\|f(x) - C(x) - Q(x) - A(x)\|_Y \leq \frac{M^3}{4k^2}\left\{\left[\tilde{\psi}_e(x) + \tilde{\psi}_e(-x)\right]^{1/p}\right\}$$
$$+ \frac{M^7}{96}\left\{4\left[\tilde{\psi}_a(x) + \tilde{\psi}_a(-x)\right]^{1/p} + \left[\tilde{\psi}_c(x) + \tilde{\psi}_c(-x)\right]^{1/p}\right\} \tag{3.111}$$

for all $x \in X$, where $\tilde{\psi}_e(x)$, $\tilde{\psi}_a(x)$, and $\tilde{\psi}_c(x)$ are defined as in Eqs. [3.44], [3.108], *and* [3.109].

Proof. Assume that $\varphi : X \times X \to [0,\infty)$ satisfies Eq. [3.43] for all $x, y \in X$ and Eq. [3.44] for all $x \in X$. Let $f_e(x) = \frac{1}{2}(f(x) + f(-x))$ for all $x \in X$, then $f_e(0) = 0$, $f_e(-x) = f_e(x)$, and

$$\|D_{f_e}(x,y)\| \leq \tilde{\varphi}(x,y)$$

for all $x, y \in X$, where $\tilde{\varphi}(x,y) := \frac{M}{2}(\varphi(x,y) + \varphi(-x,-y))$. So

$$\lim_{n\to\infty} k^{2nj}\tilde{\varphi}\left(\frac{x}{k^{nj}}, \frac{y}{k^{nj}}\right) = 0$$

for all $x, y \in X$. Since

$$\tilde{\varphi}^p(x,y) \leq \frac{M^p}{2^p}(\varphi^p(x,y) + \varphi^p(-x,-y))$$

for all $x, y \in X$, then

$$\sum_{i=\frac{1+j}{2}}^{\infty} k^{2ipj}\tilde{\varphi}^{p}\left(0, \frac{x}{k^{ij}}\right) < \infty$$

for all $x \in X$. Hence, from Theorem 3.2.6, there exist a unique quadratic function $Q : X \to Y$ such that

$$\|f_e(x) - Q(x)\|_Y \le \frac{M}{2k^2}\left[\tilde{\tilde{\psi}}_e(x)\right]^{1/p} \tag{3.112}$$

for all $x \in X$, where

$$\tilde{\tilde{\psi}}_e(x) := \sum_{i=\frac{1+j}{2}}^{\infty} k^{2ipj}\tilde{\varphi}^{p}\left(0, \frac{x}{k^{ij}}\right)$$

for all $x \in X$. It is clear that

$$\tilde{\tilde{\psi}}_e(x) \le \frac{M^p}{2^p}\left[\tilde{\psi}_q(x) + \tilde{\psi}_q(-x)\right]$$

for all $x \in X$. Therefore, it follows from Eq. [3.112] that

$$\|f_e(x) - Q(x)\|_Y \le \frac{M^2}{4k^2}\left[\tilde{\psi}_e(x) + \tilde{\psi}_e(-x)\right]^{1/p} \tag{3.113}$$

for all $x \in X$.

Also, let $f_o(x) = \frac{1}{2}(f(x) - f(-x))$ for all $x \in X$. By using the above method and Theorem 3.2.14, it follows that there exist a unique cubic function $C : X \to Y$ and a unique additive function $A : X \to Y$ such that

$$\|f_o(x)-C(x)-A(x)\|_Y \le \frac{M^6}{96}\left(4\left[\tilde{\psi}_a(x) + \tilde{\psi}_a(-x)\right]^{1/p} + \left[\tilde{\psi}_c(x) + \tilde{\psi}_c(-x)\right]^{1/p}\right) \tag{3.114}$$

for all $x \in X$. Hence Eq. [3.111] follows from Eqs. [3.113] and [3.114]. Now, if $\varphi : X \times X \to [0, \infty)$ satisfies Eq. [3.105] for all $x, y \in X$ and Eq. [3.106] for all $u \in \{x, 2x, (k-1)x, (k+1)x, (2k-1)x, (2k+1)x\}$ and all $y \in \{x, 2x, 3x\}$, we can prove the theorem by a similar technique. □

Corollary 3.2.18 (Eshaghi and Khodaei [28]). *Let θ, r, s be nonnegative real numbers such that $r, s > 3$ or $2 < r, s < 3$ or $1 < r, s < 2$ or $r, s < 1$. Suppose that a function $f : X \to Y$ satisfies the inequality* [3.55] *for all $x, y \in X$. Furthermore, assume that $f(0) = 0$ for the case f is even. Then there exist a unique cubic function $C : X \to Y$, a unique quadratic function $Q : X \to Y$ and a unique additive function $A : X \to Y$ such that*

$$\|f(x)-C(x)-Q(x)-A(x)\|_Y \leq \frac{M^7\theta}{6k^2(k^2-1)}(\lambda_a(x)+\lambda_c(x))+\frac{M^3\theta}{2}\left(\frac{1}{|k^{2p}-k^{sp}|}\|x\|_X^{sp}\right)^{1/p}$$

for all $x \in X$, where $\lambda_a(x)$ and $\lambda_c(x)$ are defined as in Corollary 3.2.15.

Proof. Put $\varphi(x,y) := \theta\left(\|x\|_X^r + \|y\|_X^s\right)$, since

$$\|D_{f_e}(x,y)\| \leq M\varphi(x,y), \quad \|D_{f_o}(x,y)\| \leq M\varphi(x,y)$$

for all $x, y \in X$. Thus the result follows from Corollaries 3.2.7 and 3.2.15. □

3.3 Mixed foursome of functional equations

Let $k \in \mathbb{Z}/\{0, \pm 1\}$. In this section, we deal with the generalized Hyers-Ulam stability of the functional equation

$$f(x+ky)+f(x-ky)=k^2(f(x+y)+f(x-y))+2(1-k^2)f(x) + \frac{k^2(k^2-1)}{12}(f(2y)+f(-2y)-4f(y)-4f(-y)) \tag{3.115}$$

derived from quartic, cubic, quadratic, and additive functions. It is easy to see that the function $f(x) = ax^4 + bx^3 + cx^2 + dx$ is a solution of the functional equation [3.115]. We investigate the general solution of functional equation [3.115] when f is a mapping between vector spaces and establish the generalized Hyers-Ulam stability of the functional equation [3.115] in non-Archimedean normed spaces. The results achieved in this section are comprehensive in that they contain the results in the papers obtained by Chang and Jung [23], Eshaghi Gordji et al. [32], Jun and Kim [24], Kim [33], Najati and Moghimi [25], Najati and Zamani Eskandani [34], and also some other papers.

In 1897, Hensel [35] introduced a normed space which does not have the Archimedean property. A non-Archimedean field is a field K equipped with a function (valuation) $|\cdot|$ from K into $[0,\infty)$ such that $|r| = 0$ if and only if $r = 0$, $|rs| = |r||s|$, and $|r+s| \leq \max\{|r|,|s|\}$ for all $r, s \in K$. Clearly $|1| = |-1| = 1$ and $|n| \leq 1$ for all $n \in N$. An example of a non-Archimedean valuation is the function $|\cdot|$ taking everything but 0 into 1 and $|0| = 0$. This valuation is called trivial.

Definition 3.3.1 ([35]). Let X be a vector space over a scalar field K with a non-Archimedean non-trivial valuation $|\cdot|$. A function $\|\cdot\| : X \to \mathbb{R}$ is a non-Archimedean norm (valuation) if it satisfies the following conditions:

(NA_1) $\|x\| = 0$ if and only if $x = 0$;

(NA_2) $\|rx\| = |r|\|x\|$ for all $r \in K$ and $x \in X$;

(NA_3) $\|x+y\| \leq \max\{\|x\|,\|y\|\}$ for all $x, y \in X$ (the strong triangle inequality).

Then $(X, \|\cdot\|)$ is called a non-Archimedean space.

Remark 3.3.2 ([35]). Thanks to the inequality

$$\|x_m - x_l\| \leq \max\{\|x_{j+1} - x_j\| : l \leq j \leq m-1\}, \quad (m > l),$$

a sequence $\{x_m\}$ is Cauchy if and only if $\{x_{m+1} - x_m\}$ converges to zero in a non-Archimedean space. By a complete non-Archimedean space we mean one in which every Cauchy sequence is convergent.

The most important examples of non-Archimedean spaces are p-adic numbers. A key property of p-adic numbers is that they do not satisfy the Archimedean axiom: for $x, y > 0$, there exists $n \in \mathbb{N}$ such that $x < ny$.

Example 3.3.3 ([35]). Let p be a prime number. For any nonzero rational number, $x = \frac{a}{b}p^{n_x}$ such that a and b are integers not divisible by p, define the p-adic absolute value $|x|_p := p^{-n_x}$. Then $|\cdot|$ is a non-Archimedean norm on $\mathbb{Q}$. The completion of $\mathbb{Q}$ with respect to $|\cdot|$ is denoted by $\mathbb{Q}_p$ which is called the p-adic number field.

Note that if $p > 3$, then $|2^n| = 1$ for each integer n. Arriola and Beyer [36] investigated the stability of approximate additive functions $f : \mathbb{Q}_p \to \mathbb{R}$. They showed that if $f : \mathbb{Q}_p \to \mathbb{R}$ is a continuous function for which there exists a fixed $\varepsilon > 0$ such that

$$|f(x+y) - f(x) - f(y)| \leq \varepsilon$$

for all $x, y \in \mathbb{Q}_p$, then there exists a unique additive function $T : \mathbb{Q}_p \to \mathbb{R}$ such that

$$|f(x) - T(x)| \leq \varepsilon$$

for all $x \in \mathbb{Q}_p$.

Throughout this section, assume that G is an additive group and X is a complete non-Archimedean space. Before taking up the main subject, given $f : G \times G \to X$, we define the difference operator

$$\begin{aligned} Df(x,y) = & f(x+ky) + f(x-ky) - k^2 f(x+y) - k^2 f(x-y) \\ & - (k^2-1)\left(\frac{k^2}{12}(\tilde{f}(2y) - 4\tilde{f}(y)) - 2f(x)\right) \end{aligned}$$

where $\tilde{f}(y) := f(y) + f(-y)$ and $k \in \mathbb{Z}\backslash\{0, \pm 1\}$, for all $x, y \in G$.

Theorem 3.3.4 (Eshaghi et al. [37]). *Let $\ell \in \{1, -1\}$ be fixed and let $\varphi : G \times G \to [0, \infty)$ be a function such that*

$$\lim_{n\to\infty} |2|^{n\ell} \varphi\left(\frac{x}{2^{n\ell}}, \frac{y}{2^{n\ell}}\right) = 0 = \lim_{n\to\infty} |2|^{n\ell} \tilde{\psi}\left(\frac{x}{2^{n\ell}}\right) \tag{3.116}$$

for all $x, y \in G$, and

$$\psi_a(x) = \lim_{n\to\infty} \max\left\{|2|^{\ell(j+\frac{1+\ell}{2})} \tilde{\psi}\left(\frac{x}{2^{\ell(j+\frac{1+\ell}{2})}}\right) : 0 \leq j < n\right\} \tag{3.117}$$

for all $x \in G$ exists, where

$$\begin{aligned}\tilde{\psi}(x) := \frac{1}{|k^2(k^2-1)|} & \max\{|2| \max\{|2(k^2-1)|\varphi(x,x), \\ & \max\{|k^2|\varphi(2x,x), \varphi(x,2x)\}, \max\{\varphi((k+1)x,x), \varphi((k-1)x,x)\}\}, \\ & \max\{\max\{\varphi(x,x), |k^2|\varphi(2x,2x)\}, \max\{|2(k^2-1)|\varphi(x,2x), \varphi(x,3x)\}, \\ & \max\{\varphi((2k+1)x,x), \varphi((2k-1)x,x)\}\}\}\end{aligned} \tag{3.118}$$

for all $x \in G$. Suppose that an odd function $f : G \to X$ satisfies the inequality

$$\|Df(x,y)\| \leq \varphi(x,y) \tag{3.119}$$

for all $x, y \in G$. Then there exists a additive function $A : G \to X$ such that

$$\|f(2x) - 8f(x) - A(x)\| \leq \frac{1}{|2|}\psi_a(x) \tag{3.120}$$

for all $x \in G$; if

$$\lim_{i\to\infty} \lim_{n\to\infty} \max\left\{|2|^{\ell\left(j+\frac{1+\ell}{2}\right)} \tilde{\psi}\left(\frac{x}{2^{\ell\left(j+\frac{1+\ell}{2}\right)}}\right) : i \leq j < n+i\right\} = 0, \tag{3.121}$$

then A is the unique additive function satisfying Eq. [3.120].

Proof. Let $\ell = 1$. It follows from Eq. [3.119] and oddness of f that

$$\|f(ky+x) - f(ky-x) - k^2 f(x+y) - k^2 f(x-y) + 2(k^2-1)f(x)\| \leq \varphi(x,y) \tag{3.122}$$

for all $x, y \in G$. Putting $y = x$ in Eq. [3.122], we have

$$\|f((k+1)x) - f((k-1)x) - k^2 f(2x) + 2(k^2-1)f(x)\| \leq \varphi(x,x) \tag{3.123}$$

for all $x \in G$. It follows from Eq. [3.123] that

$$\|f(2(k+1)x) - f(2(k-1)x) - k^2 f(4x) + 2(k^2-1)f(2x)\| \leq \varphi(2x,2x) \tag{3.124}$$

for all $x \in G$. Replacing x and y by $2x$ and x in Eq. [3.122], respectively, we get

$$\|f((k+2)x) - f((k-2)x) - k^2 f(3x) - k^2 f(x) + 2(k^2-1)f(2x)\| \leq \varphi(2x,x) \tag{3.125}$$

for all $x \in G$. Setting $y = 2x$ in Eq. [3.122], one obtains

$$\|f((2k+1)x) - f((2k-1)x) - k^2 f(3x) - k^2 f(-x) + 2(k^2-1)f(x)\| \leq \varphi(x,2x) \tag{3.126}$$

for all $x \in G$. Putting $y = 3x$ in Eq. [3.122], we obtain

$$\|f((3k+1)x) - f((3k-1)x) - k^2 f(4x) - k^2 f(-2x) + 2(k^2-1)f(x)\| \leq \varphi(x, 3x) \tag{3.127}$$

for all $x \in G$. Replacing x and y by $(k+1)x$ and x in Eq. [3.122], respectively, we get

$$\begin{aligned} &\|f((2k+1)x) - f(-x) - k^2 f((k+2)x) - k^2 f(kx) + 2(k^2-1)f((k+1)x)\| \\ &\quad \leq \varphi((k+1)x, x) \end{aligned} \tag{3.128}$$

for all $x \in G$. Replacing x and y by $(k-1)x$ and x in Eq. [3.122], respectively, one gets

$$\begin{aligned} &\|f((2k-1)x) - f(x) - k^2 f((k-2)x) - k^2 f(kx) + 2(k^2-1)f((k-1)x)\| \\ &\quad \leq \varphi((k-1)x, x) \end{aligned} \tag{3.129}$$

for all $x \in G$. Replacing x and y by $(2k+1)x$ and x in Eq. [3.122], respectively, we obtain

$$\begin{aligned} &\|f((3k+1)x) - f(-(k+1)x) - k^2 f(2(k+1)x) - k^2 f(2kx) \\ &\quad + 2(k^2-1)f((2k+1)x)\| \leq \varphi((2k+1)x, x) \end{aligned} \tag{3.130}$$

for all $x \in G$. If we replace x and y by $(2k-1)x$ and x in Eq. [3.122], respectively, then we have

$$\begin{aligned} &\|f((3k-1)x) - f(-(k-1)x) - k^2 f(2(k-1)x) - k^2 f(2kx) \\ &\quad + 2(k^2-1)f((2k-1)x)\| \leq \varphi((2k-1)x, x) \end{aligned} \tag{3.131}$$

for all $x \in G$. It follows from Eqs. [3.123], [3.125], [3.126], [3.128], and [3.129] that

$$\begin{aligned} \|f(3x) - 4f(2x) + 5f(x)\| \leq \frac{1}{|k^2(k^2-1)|} &\max\{|2(k^2-1)|\varphi(x,x), \\ &\max\{|k^2|\varphi(2x,x), \varphi(x,2x)\}, \\ &\max\{\varphi((k+1)x,x), \varphi((k-1)x,x)\}\} \end{aligned} \tag{3.132}$$

for all $x \in G$. Also, from Eqs. [3.123], [3.124], [3.126], [3.127], [3.130], and [3.131], we conclude that

$$\begin{aligned} &\|f(4x) - 2f(3x) - 2f(2x) + 6f(x)\| \\ &\leq \frac{1}{|k^2(k^2-1)|} \max\{\max\{\varphi(x,x), |k^2|\varphi(2x,2x)\}, \\ &\quad \max\{|2(k^2-1)|\varphi(x,2x), \varphi(x,3x)\}, \max\{\varphi((2k+1)x,x), \varphi((2k-1)x,x)\}\} \end{aligned} \tag{3.133}$$

for all $x \in G$. Finally, by using Eqs. [3.132] and [3.133], we obtain that

$$\begin{aligned}
&\|f(4x) - 10f(2x) + 16f(x)\| \\
&\leq \frac{1}{|k^2(k^2-1)|} \max\{|2| \max\{|2(k^2-1)|\varphi(x,x), \\
&\quad \max\{|k^2|\varphi(2x,x), \varphi(x,2x)\}, \max\{\varphi((k+1)x,x), \varphi((k-1)x,x)\}\}, \\
&\quad \max\{\max\{\varphi(x,x), |k^2|\varphi(2x,2x)\}, \max\{|2(k^2-1)|\varphi(x,2x), \varphi(x,3x)\}, \\
&\quad \max\{\varphi((2k+1)x,x), \varphi((2k-1)x,x)\}\}\}
\end{aligned} \tag{3.134}$$

for all $x \in G$, and let

$$\begin{aligned}
\tilde{\psi}(x) := &\frac{1}{|k^2(k^2-1)|} \max\{|2| \max\{|2(k^2-1)|\varphi(x,x), \\
&\max\{|k^2|\varphi(2x,x), \varphi(x,2x)\}, \max\{\varphi((k+1)x,x), \varphi((k-1)x,x)\}\}, \\
&\max\{\max\{\varphi(x,x), |k^2|\varphi(2x,2x)\}, \max\{|2(k^2-1)|\varphi(x,2x), \varphi(x,3x)\}, \\
&\max\{\varphi((2k+1)x,x), \varphi((2k-1)x,x)\}\}\}
\end{aligned} \tag{3.135}$$

for all $x \in G$. Thus Eq. [3.134] means that

$$\|f(4x) - 10f(2x) + 16f(x)\| \leq \tilde{\psi}(x) \tag{3.136}$$

for all $x \in G$. Let $g_1 : G \to X$ be a function defined by $g_1(x) := f(2x) - 8f(x)$ for all $x \in G$. From Eq. [3.136], we conclude that

$$\|g_1(2x) - 2g_1(x)\| \leq \tilde{\psi}(x) \tag{3.137}$$

for all $x \in G$. If we replace x in Eq. [3.137] by $\frac{x}{2^{n+1}}$, we get

$$\left\|2^{n+1}g_1\left(\frac{x}{2^{n+1}}\right) - 2^n g_1\left(\frac{x}{2^n}\right)\right\| \leq |2|^n \tilde{\psi}\left(\frac{x}{2^{n+1}}\right) \tag{3.138}$$

for all $x \in G$. It follows from Eqs. [3.116] and [3.138] that the sequence $\{2^n g_1(\frac{x}{2^n})\}$ is Cauchy. Since X is complete, we conclude that $\{2^n g_1(\frac{x}{2^n})\}$ is convergent. So one can define the function $A : G \to X$ by

$$A(x) := \lim_{n\to\infty} 2^n g_1\left(\frac{x}{2^n}\right) \tag{3.139}$$

for all $x \in G$. It follows from Eqs. [3.137] and [3.138] by using induction that

$$\left\|g_1(x) - 2^n g_1\left(\frac{x}{2^n}\right)\right\| \leq \frac{1}{|2|} \max\left\{|2|^{j+1}\tilde{\psi}\left(\frac{x}{2^{j+1}}\right) : 0 \leq j < n\right\} \tag{3.140}$$

for all $n \in \mathbb{N}$ and all $x \in G$. By taking n to approach infinity in Eq. [3.140] and using Eq. [3.117], one gets Eq. [3.120]. Now we show that A is additive. It follows from Eqs. [3.116], [3.138], and [3.139] that

$$\begin{aligned}\|A(2x) - 2A(x)\| &= \lim_{n\to\infty} \left\|2^n g_1\left(\frac{x}{2^{n-1}}\right) - 2^{n+1} g_1\left(\frac{x}{2^n}\right)\right\| \\ &= |2| \lim_{n\to\infty} \left\|2^{n-1} g_1\left(\frac{x}{2^{n-1}}\right) - 2^n g_1\left(\frac{x}{2^n}\right)\right\| \\ &\le \lim_{n\to\infty} |2|^n \tilde{\psi}\left(\frac{x}{2^n}\right) = 0\end{aligned}$$

for all $x \in G$. So

$$A(2x) = 2A(x) \tag{3.141}$$

for all $x \in G$. On the other hand, it follows from Eqs. [3.116], [3.119], and [3.139] that

$$\begin{aligned}\|DA(x,y)\| &= \lim_{n\to\infty} |2|^n \left\|Dg_1\left(\frac{x}{2^n}, \frac{y}{2^n}\right)\right\| \\ &= \lim_{n\to\infty} |2|^n \left\|Df\left(\frac{x}{2^{n-1}}, \frac{y}{2^{n-1}}\right) - 8Df\left(\frac{x}{2^n}, \frac{y}{2^n}\right)\right\| \\ &\le \lim_{n\to\infty} |2|^n \max\left\{\varphi\left(\frac{x}{2^{n-1}}, \frac{y}{2^{n-1}}\right), |8|\varphi\left(\frac{x}{2^n}, \frac{y}{2^n}\right)\right\} = 0\end{aligned}$$

for all $x, y \in G$. Hence the function A satisfies Eq. [3.115]. Thus, the function $x \rightsquigarrow A(2x) - 8A(x)$ is cubic-additive. Therefore, Eq. [3.141] implies that the function A is additive. If A' is another additive function satisfying Eq. [3.120], then

$$\begin{aligned}\|A(x) - A'(x)\| &= \lim_{i\to\infty} |2|^i \left\|A\left(\frac{x}{2^i}\right) - A'\left(\frac{x}{2^i}\right)\right\| \\ &\le \lim_{i\to\infty} |2|^i \max\left\{\left\|A\left(\frac{x}{2^i}\right) - g_1\left(\frac{x}{2^i}\right)\right\|, \left\|g_1\left(\frac{x}{2^i}\right) - A'\left(\frac{x}{2^i}\right)\right\|\right\} \\ &\le \frac{1}{|2|} \lim_{i\to\infty} \lim_{n\to\infty} \max\left\{|2|^{j+1}\tilde{\psi}\left(\frac{x}{2^{j+1}}\right) : i \le j < n+i\right\} = 0\end{aligned}$$

for all $x \in G$. Therefore $A = A'$. For $\ell = -1$, we can prove the theorem by a similar technique. □

Theorem 3.3.5 (Eshaghi et al. [37]). *Let $\ell \in \{1, -1\}$ be fixed and let $\varphi : G \times G \to [0, \infty)$ be a function such that*

$$\lim_{n\to\infty} |2|^{3n\ell} \varphi\left(\frac{x}{2^{n\ell}}, \frac{y}{2^{n\ell}}\right) = 0 = \lim_{n\to\infty} |2|^{3n\ell} \tilde{\psi}\left(\frac{x}{2^{n\ell}}\right) \tag{3.142}$$

for all $x, y \in G$, and

$$\psi_c(x) = \lim_{n\to\infty} \max\left\{|2|^{3\ell\left(j+\frac{1+\ell}{2}\right)} \tilde{\psi}\left(\frac{x}{2^{\ell\left(j+\frac{1+\ell}{2}\right)}}\right) : 0 \le j < n\right\} \tag{3.143}$$

for all $x \in G$ exists, where $\tilde{\psi}(x)$ is defined as in Eq. [3.118] for all $x \in G$. Suppose that an odd function $f : G \to X$ satisfies the inequality [3.119] *for all $x, y \in G$. Then there exists a cubic function $C : G \to X$ such that*

$$\|f(2x) - 2f(x) - C(x)\| \leq \frac{1}{|2|^3}\psi_c(x) \tag{3.144}$$

for all $x \in G$; if

$$\lim_{i\to\infty}\lim_{n\to\infty}\max\left\{|2|^{3\ell\left(j+\frac{1+\ell}{2}\right)}\tilde{\psi}\left(\frac{x}{2^{\ell\left(j+\frac{1+\ell}{2}\right)}}\right) : i \leq j < n+i\right\} = 0, \tag{3.145}$$

then C is the unique cubic function satisfying Eq. [3.144].

Proof. Let $\ell = -1$. Similar to the proof of Theorem 3.3.4, we have

$$\|f(4x) - 10f(2x) + 16f(x)\| \leq \tilde{\psi}(x) \tag{3.146}$$

for all $x \in G$, where $\tilde{\psi}(x)$ is defined as in Eq. [3.118] for all $x \in G$. Let $g_2 : G \to X$ be a function defined by $g_2(x) := f(2x) - 2f(x)$ for all $x \in G$. From Eq. [3.146], we conclude that

$$\|g_2(2x) - 8g_2(x)\| \leq \tilde{\psi}(x) \tag{3.147}$$

for all $x \in G$. If we replace x in Eq. [3.147] by $2^{n-1}x$, we get

$$\left\|\frac{g_2(2^n x)}{2^{3n}} - \frac{g_2(2^{n-1}x)}{2^{3(n-1)}}\right\| \leq \frac{1}{|2|^{3n}}\tilde{\psi}(2^{n-1}x) \tag{3.148}$$

for all $x \in G$. It follows from Eqs. [3.142] and [3.148] that the sequence $\{\frac{g_2(2^n x)}{2^{3n}}\}$ is Cauchy. Since X is complete, we conclude that $\{\frac{g_2(2^n x)}{2^{3n}}\}$ is convergent. So one can define the function $C : G \to X$ by

$$C(x) := \lim_{n\to\infty}\frac{g_2(2^n x)}{2^{3n}} \tag{3.149}$$

for all $x \in G$. It follows from Eqs. [3.147] and [3.148] by using induction that

$$\left\|g_2(x) - \frac{g_2(2^n x)}{2^{3n}}\right\| \leq \frac{1}{|2|^3}\max\left\{\frac{1}{|2|^{3j}}\tilde{\psi}(2^j x) : 0 \leq j < n\right\} \tag{3.150}$$

for all $n \in \mathbb{N}$ and all $x \in G$. By taking n to approach infinity in Eq. [3.150] and using Eq. [3.143], one gets Eq. [3.144]. Now we show that C is cubic. It follows from Eqs. [3.142], [3.148], and [3.149] that

$$\begin{aligned}\|C(2x)-8C(x)\| &= \lim_{n\to\infty}\left\|\frac{g_2(2^{n+1}x)}{2^{3n}}-\frac{2^3 g_2(2^n x)}{2^{3n}}\right\| \\ &= |2|^3\lim_{n\to\infty}\left\|\frac{g_2(2^{n+1}x)}{2^{3(n+1)}}-\frac{g_2(2^n x)}{2^{3n}}\right\| \\ &\le \lim_{n\to\infty}\frac{1}{|2|^{3n}}\tilde{\psi}(2^n x)=0\end{aligned}$$

for all $x \in G$. So

$$C(2x) = 8C(x) \tag{3.151}$$

for all $x \in G$. On the other hand, it follows from Eqs. [3.119], [3.142], and [3.149] that

$$\begin{aligned}\|DC(x,y)\| &= \lim_{n\to\infty}\frac{1}{|2|^{3n}}\|Dg_2(2^n x, 2^n y)\| \\ &= \lim_{n\to\infty}\frac{1}{|2|^{3n}}\|Df(2^{n+1}x, 2^{n+1}y)-2Df(2^n x, 2^n y)\| \\ &\le \lim_{n\to\infty}\frac{1}{|2|^{3n}}\max\{\varphi(2^{n+1}x, 2^{n+1}y), |2|\varphi(2^n x, 2^n y)\}=0\end{aligned}$$

for all $x, y \in G$. Hence the function C satisfies Eq. [3.115]. Thus, the function $x \mapsto C(2x) - 2C(x)$ is cubic-additive. Therefore Eq. [3.151] implies that the function C is cubic. The rest of the proof is similar to the proof of Theorem 3.3.4. For $\ell = 1$, we can prove the theorem by a similar technique. □

Theorem 3.3.6 (Eshaghi et al. [37]). *Let $\ell \in \{1, -1\}$ be fixed and let $\varphi : G \times G \to [0, \infty)$ be a function such that*

$$\begin{aligned}&\lim_{n\to\infty}\left\{\left(\frac{1-\ell}{2}\right)|2|^{n\ell}\varphi\left(\frac{x}{2^{n\ell}}, \frac{y}{2^{n\ell}}\right)+\left(\frac{1+\ell}{2}\right)|2|^{3n\ell}\varphi\left(\frac{x}{2^{n\ell}}, \frac{y}{2^{n\ell}}\right)\right\} \\ &\quad = 0 = \lim_{n\to\infty}\left\{\left(\frac{1-\ell}{2}\right)|2|^{n\ell}\tilde{\psi}\left(\frac{x}{2^{n\ell}}\right)+\left(\frac{1+\ell}{2}\right)|2|^{3n\ell}\tilde{\psi}\left(\frac{x}{2^{n\ell}}\right)\right\}\end{aligned} \tag{3.152}$$

for all $x, y \in G$, and

$$\begin{aligned}\lim_{n\to\infty}\max\Bigg\{&\Bigg[\left(\frac{1-\ell}{2}\right)|2|^{\ell(j+\frac{1+\ell}{2})} \\ &+\left(\frac{1+\ell}{2}\right)|2|^{3\ell(j+\frac{1+\ell}{2})}\Bigg]\tilde{\psi}\left(\frac{x}{2^{\ell(j+\frac{1+\ell}{2})}}\right) : 0 \le j < n\Bigg\}\end{aligned} \tag{3.153}$$

for all $x \in G$ exists, where $\tilde{\psi}(x)$ is defined as in Eq. [3.118] for all $x \in G$. Suppose that an odd function $f : G \to X$ satisfies the inequality [3.119] *for all $x, y \in G$. Then there exist an additive function $A : G \to X$ and a cubic function $C : G \to X$ such that*

$$\|f(x) - A(x) - C(x)\| \leq \frac{1}{|12|} \max \left\{ \psi_a(x), \frac{1}{|4|} \psi_c(x) \right\} \tag{3.154}$$

for all $x \in G$, where $\psi_a(x)$ and $\psi_c(x)$ are defined as in Theorems 3.3.4 and 3.3.5. Moreover, if

$$\begin{aligned} \lim_{i\to\infty} \lim_{n\to\infty} \max \Bigg\{ \left[\left(\frac{1-\ell}{2}\right) |2|^{\ell\left(j+\frac{1+\ell}{2}\right)} + \left(\frac{1+\ell}{2}\right) |2|^{3\ell\left(j+\frac{1+\ell}{2}\right)} \right] \\ \cdot \tilde{\psi}\left(\frac{x}{2^{\ell\left(j+\frac{1+\ell}{2}\right)}}\right) : i \leq j < n+i \Bigg\} = 0, \end{aligned} \tag{3.155}$$

then A is the unique additive function and C is the unique cubic function satisfying Eq. [3.154].

Proof. Let $\ell = 1$. By Theorems 3.3.4 and 3.3.5, there exist an additive function $A_0 : G \to X$ and a cubic function $C_0 : G \to X$ such that

$$\|f(2x) - 8f(x) - A_0(x)\| \leq \frac{1}{|2|} \psi_a(x),$$
$$\|f(2x) - 2f(x) - C_0(x)\| \leq \frac{1}{|2|^3} \psi_c(x)$$

for all $x \in G$. So we obtain Eq. [3.152] by letting $A(x) = -1/6A_0(x)$ and $C(x) = 1/6C_0(x)$ for all $x \in G$.

To prove the uniqueness property of A and C, let $C', A' : G \to X$ be another additive and cubic functions satisfying Eq. [3.152]. Let $\overline{A} = A - A'$ and $\overline{C} = C - C'$. Hence,

$$\begin{aligned} \|\overline{A}(x) + \overline{C}(x)\| &\leq \max\{\|f(x) - A(x) - C(x)\|, \|f(x) - A'(x) - C'(x)\|\} \\ &\leq \frac{1}{|12|} \max \left\{ \psi_a(x), \frac{1}{|4|} \psi_c(x) \right\} \end{aligned}$$

for all $x \in G$. Since

$$\begin{aligned} &\lim_{i\to\infty} \lim_{n\to\infty} \max \left\{ |2|^{(j+1)} \tilde{\psi}\left(\frac{x}{2^{j+1}}\right) : i \leq j < n+i \right\} \\ &\quad = 0 = \lim_{i\to\infty} \lim_{n\to\infty} \max \left\{ |2|^{3(j+1)} \tilde{\psi}\left(\frac{x}{2^{j+1}}\right) : i \leq j < n+i \right\}, \end{aligned}$$

so

$$\lim_{n\to\infty} |2|^{3n} \left\| \overline{A}\left(\frac{x}{2^n}\right) + \overline{C}\left(\frac{x}{2^n}\right) \right\| = 0$$

for all $x \in X$. Therefore, we get $\overline{C} = 0$ and then $\overline{A} = 0$, and the proof is complete. For $\ell = -1$, we can prove the theorem by a similar technique. □

Theorem 3.3.7 (Eshaghi et al. [37]). *Let $\varphi : G \times G \to [0, \infty)$ be a function such that*

$$\begin{aligned} &\lim_{n\to\infty} |2|^n \varphi\left(\frac{x}{2^n}, \frac{y}{2^n}\right) = 0 = \lim_{n\to\infty} |2|^n \tilde{\psi}\left(\frac{x}{2^n}\right), \\ &\lim_{n\to\infty} \frac{1}{|2|^{3n}} \varphi\left(2^n x, 2^n y\right) = 0 = \lim_{n\to\infty} \frac{1}{|2|^{3n}} \tilde{\psi}(2^n x) \end{aligned} \tag{3.156}$$

for all $x, y \in G$, and the limits

$$\lim_{n\to\infty} \max\left\{|2|^{j+1} \tilde{\psi}\left(\frac{x}{2^{j+1}}\right) : 0 \le j < n\right\}, \quad \lim_{n\to\infty} \max\left\{\frac{1}{|2|^{3j}} \tilde{\psi}(2^j x) : 0 \le j < n\right\}$$

for all $x \in G$ exist, where $\tilde{\psi}(x)$ is defined as in Eq. [3.118] for all $x \in G$. Suppose that an odd function $f : G \to X$ satisfies the inequality [3.119] *for all $x, y \in G$. Then there exists an additive function $A : G \to X$ and a cubic function $C : G \to X$ such that*

$$\|f(x) - A(x) - C(x)\| \le \frac{1}{|12|} \max\left\{\psi_a(x), \frac{1}{|4|}\psi_c(x)\right\} \tag{3.157}$$

for all $x \in G$, where $\psi_a(x)$, and $\psi_c(x)$ are defined as in Theorems 3.3.4 and 3.3.5. Moreover, if

$$\begin{aligned} &\lim_{i\to\infty} \lim_{n\to\infty} \max\left\{|2|^{j+1} \tilde{\psi}\left(\frac{x}{2^{j+1}}\right) : i \le j < n+i\right\} \\ &\quad = 0 = \lim_{i\to\infty} \lim_{n\to\infty} \max\left\{\frac{1}{|2|^{3j}} \tilde{\psi}(2^j x) : i \le j < n+i\right\}, \end{aligned}$$

then A is the unique additive function and C is the unique cubic function satisfying Eq. [3.157].

Proof. The proof is similar to the proof of Theorem 3.3.6, and the result follows from Theorems 3.3.4 and 3.3.5. □

Theorem 3.3.8 (Eshaghi et al. [37]). *Let $\ell \in \{1, -1\}$ be fixed and let $\varphi : G \times G \to [0, \infty)$ be a function such that*

$$\lim_{n\to\infty} |2|^{2n\ell} \varphi\left(\frac{x}{2^{n\ell}}, \frac{y}{2^{n\ell}}\right) = 0 = \lim_{n\to\infty} |2|^{2n\ell} \tilde{\varphi}\left(\frac{x}{2^{n\ell}}\right) \tag{3.158}$$

for all $x, y \in G$, and the limit

$$\psi_q(x) = \lim_{n\to\infty} \max\left\{|2|^{2\ell\left(j+\frac{1+\ell}{2}\right)} \tilde{\varphi}\left(\frac{x}{2^{\ell\left(j+\frac{1+\ell}{2}\right)}}\right) : 0 \le j < n\right\} \tag{3.159}$$

for all $x \in G$ exists, where

$$\begin{aligned} \tilde{\varphi}(x) := \frac{1}{|k^2(k^2-1)|} \max\{ &\max\{|12k^2|\varphi(x,x), |12(k^2-1)|\varphi(0,x)\}, \\ &\max\{|6|\varphi(0,2x), |12|\varphi(kx,x)\}\} \end{aligned} \tag{3.160}$$

for all $x \in G$. Suppose that an even function $f : G \to X$ with $f(0) = 0$ satisfies in the inequality [3.119] *for all $x, y \in G$. Then there exists a quadratic function $Q : G \to X$ such that*

$$\|f(2x) - 16f(x) - Q(x)\| \leq \frac{1}{|2|^2}\psi_q(x) \tag{3.161}$$

for all $x \in G$. If

$$\lim_{i\to\infty}\lim_{n\to\infty}\max\left\{|2|^{2\ell(j+\frac{1+\ell}{2})}\tilde{\varphi}\left(\frac{x}{2^{\ell(j+\frac{1+\ell}{2})}}\right) : i \leq j < n+i\right\} = 0, \tag{3.162}$$

then Q is the unique quadratic function satisfying Eq. [3.161].

Proof. Let $\ell = 1$. It follows from Eq. [3.119] and using the evenness of f that

$$\begin{aligned}&\|f(x+ky) + f(x-ky) - k^2f(x+y) - k^2f(x-y) - 2(1-k^2)f(x)\\&\quad - \frac{k^2(k^2-1)}{6}(f(2y) - 4f(y))\| \leq \varphi(x,y)\end{aligned} \tag{3.163}$$

for all $x, y \in G$. Interchange x with y in Eq. [3.163], we get by the evenness of f

$$\begin{aligned}&\Big\|f(kx+y) + f(kx-y) - k^2f(x+y) - k^2f(x-y) + 2(k^2-1)f(y)\\&\quad - \frac{k^2(k^2-1)}{6}(f(2x) - 4f(x))\Big\| \leq \varphi(y,x)\end{aligned} \tag{3.164}$$

for all $x, y \in G$. Setting $y = 0$ in Eq. [3.164], we have

$$\left\|2f(kx) - 2k^2f(x) - \frac{k^2(k^2-1)}{6}(f(2x) - 4f(x))\right\| \leq \varphi(0,x) \tag{3.165}$$

for all $x \in G$. Putting $y = x$ in Eq. [3.164], we obtain

$$\begin{aligned}&\Big\|f((k+1)x) + f((k-1)x) - k^2f(2x) + 2(k^2-1)f(x)\\&\quad - \frac{k^2(k^2-1)}{6}(f(2x) - 4f(x))\Big\| \leq \varphi(x,x)\end{aligned} \tag{3.166}$$

for all $x \in G$. Replacing x and y by $2x$ and 0 in Eq. [3.164], respectively, we see that

$$\left\|2f(2kx) - 2k^2f(2x) - \frac{k^2(k^2-1)}{6}(f(4x) - 4f(2x))\right\| \leq \varphi(0,2x) \tag{3.167}$$

for all $x \in G$. Setting $y = kx$ in Eq. [3.164] and using the evenness of f, we get

$$\begin{aligned} &\Big\| f(2kx) - k^2 f((k+1)x) - k^2 f((k-1)x) + 2(k^2-1) f(kx) \\ &\quad - \frac{k^2(k^2-1)}{6}(f(2x) - 4f(x)) \Big\| \le \varphi(kx, x) \end{aligned} \tag{3.168}$$

for all $x \in G$. It follows from Eqs. [3.165], [3.166], [3.167], and [3.168] that

$$\begin{aligned} &\|f(4x) - 20f(2x) + 64f(x)\| \\ &\le \frac{1}{|k^2(k^2-1)|} \max\{\max\{|12k^2|\varphi(x,x), |12(k^2-1)|\varphi(0,x)\}, \\ &\quad \max\{|6|\varphi(0,2x), |12|\varphi(kx,x)\}\} \end{aligned} \tag{3.169}$$

for all $x \in G$. Let

$$\begin{aligned} \tilde{\varphi}(x) := \frac{1}{|k^2(k^2-1)|} \max\{ &\max\{|12k^2|\varphi(x,x), |12(k^2-1)|\varphi(0,x)\}, \\ &\max\{|6|\varphi(0,2x), |12|\varphi(kx,x)\}\} \end{aligned} \tag{3.170}$$

for all $x \in G$. Then Eq. [3.169] means

$$\|f(4x) - 20f(2x) + 64f(x)\| \le \tilde{\varphi}(x) \tag{3.171}$$

for all $x \in G$. Let $h_1 : G \to X$ be a function defined by $h_1(x) := f(2x) - 16f(x)$ for all $x \in G$. From Eq. [3.171], we conclude that

$$\|h_1(2x) - 4h_1(x)\| \le \tilde{\varphi}(x) \tag{3.172}$$

for all $x \in G$. Replacing x by $\frac{x}{2^{n+1}}$ in Eq. [3.172], we have

$$\left\| 2^{2(n+1)} h_1\left(\frac{x}{2^{n+1}}\right) - 2^{2n} h_1\left(\frac{x}{2^n}\right) \right\| \le |2|^{2n} \tilde{\varphi}\left(\frac{x}{2^{n+1}}\right) \tag{3.173}$$

for all $x \in G$. It follows from Eqs. [3.158] and [3.173] that the sequence $\{2^{2n} h_1(\frac{x}{2^n})\}$ is Cauchy. Since X is complete, we conclude that $\{2^{2n} h_1(\frac{x}{2^n})\}$ is convergent. So one can define the function $Q : G \to X$ by

$$Q(x) := \lim_{n \to \infty} 2^{2n} h_1\left(\frac{x}{2^n}\right) \tag{3.174}$$

for all $x \in G$. It follows from Eqs. [3.172] and [3.173] by using induction that

$$\left\| h_1(x) - 2^{2n} h_1\left(\frac{x}{2^n}\right) \right\| \le \frac{1}{|2|^2} \max\left\{ |2|^{2(j+1)} \tilde{\varphi}\left(\frac{x}{2^{j+1}}\right) : \ 0 \le j < n \right\} \tag{3.175}$$

for all $n \in \mathbb{N}$ and all $x \in G$. By taking n to approach infinity in Eq. [3.175] and using Eq. [3.159], one gets Eq. [3.161]. Now we show that Q is quadratic. It follows from Eqs. [3.158], [3.173], and [3.174] that

$$\begin{aligned}\|Q(2x) - 4Q(x)\| &= \lim_{n\to\infty} \left\| 2^{2n} h_1\left(\frac{x}{2^{n-1}}\right) - 2^{2(n+1)} h_1\left(\frac{x}{2^n}\right)\right\| \\ &= \lim_{n\to\infty} |2|^2 \left\| 2^{2(n-1)} h_1\left(\frac{x}{2^{n-1}}\right) - 2^{2n} h_1\left(\frac{x}{2^n}\right)\right\| \\ &\leq \lim_{n\to\infty} |2|^{2(n+1)} \tilde{\varphi}\left(\frac{x}{2^{n+1}}\right) = 0\end{aligned}$$

for all $x \in G$. So

$$Q(2x) = 4Q(x) \tag{3.176}$$

for all $x \in G$. On the other hand it follows from Eqs. [3.119], [3.158], and [3.174] that

$$\begin{aligned}\|DQ(x,y)\| &= \lim_{n\to\infty} |2|^{2n} \left\| Dh_1\left(\frac{x}{2^n}, \frac{y}{2^n}\right)\right\| \\ &= \lim_{n\to\infty} |2|^{2n} \left\| Df\left(\frac{x}{2^{n-1}}, \frac{y}{2^{n-1}}\right) - 16Df\left(\frac{x}{2^n}, \frac{y}{2^n}\right)\right\| \\ &\leq \lim_{n\to\infty} |2|^{2n} \max\left\{\varphi\left(\frac{x}{2^{n-1}}, \frac{y}{2^{n-1}}\right), |16|\varphi\left(\frac{x}{2^n}, \frac{y}{2^n}\right)\right\} = 0\end{aligned}$$

for all $x, y \in G$. Hence the function Q satisfies Eq. [3.115]. Thus, the function $x \mapsto Q(2x) - 16Q(x)$ is quartic-quadratic. Now, Eq. [3.176] implies that the function Q is quadratic. The rest of the proof is similar to the proof of Theorem 3.3.4. For $\ell = -1$, we can prove the theorem by a similar technique. □

Theorem 3.3.9 (Eshaghi et al. [37]). *Let $\ell \in \{1, -1\}$ be fixed and let $\varphi : G \times G \to [0, \infty)$ be a function such that*

$$\lim_{n\to\infty} |2|^{4n\ell} \varphi\left(\frac{x}{2^{n\ell}}, \frac{y}{2^{n\ell}}\right) = 0 = \lim_{n\to\infty} |2|^{4n\ell} \tilde{\varphi}\left(\frac{x}{2^{n\ell}}\right) \tag{3.177}$$

for all $x, y \in G$, and

$$\psi_v(x) = \lim_{n\to\infty} \max\left\{ |2|^{4\ell\left(j+\frac{1+\ell}{2}\right)} \tilde{\varphi}\left(\frac{x}{2^{\ell\left(j+\frac{1+\ell}{2}\right)}}\right) : 0 \leq j < n\right\} \tag{3.178}$$

exists for all $x \in G$, where $\tilde{\varphi}(x)$ is defined as in Eq. [3.160] for all $x \in G$. Suppose that an even function $f : G \to X$ with $f(0) = 0$ satisfies in the inequality [3.119] *for all $x, y \in G$. Then there exist a quartic function $V : G \to X$ such that*

$$\|f(2x) - 4f(x) - V(x)\| \leq \frac{1}{|2|^4} \psi_v(x) \tag{3.179}$$

for all $x \in G$. If

$$\lim_{i\to\infty}\lim_{n\to\infty}\max\left\{|2|^{4\ell\left(j+\frac{1+\ell}{2}\right)}\tilde{\varphi}\left(\frac{x}{2^{\ell\left(j+\frac{1+\ell}{2}\right)}}\right) : 0 \le j < n\right\} = 0, \tag{3.180}$$

then V is the unique quartic function satisfying Eq. [3.179].

Theorem 3.3.10 (Eshaghi et al. [37]). *Let $\ell \in \{1, -1\}$ be fixed and let $\varphi : G \times G \to [0, \infty)$ be a function such that*

$$\begin{aligned}\lim_{n\to\infty}&\left\{\left(\frac{1-\ell}{2}\right)|2|^{2n\ell}\varphi\left(\frac{x}{2^{n\ell}}, \frac{y}{2^{n\ell}}\right) + \left(\frac{1+\ell}{2}\right)|2|^{4n\ell}\varphi\left(\frac{x}{2^{n\ell}}, \frac{y}{2^{n\ell}}\right)\right\}\\ &= 0 = \lim_{n\to\infty}\left\{\left(\frac{1-\ell}{2}\right)|2|^{2n\ell}\tilde{\varphi}\left(\frac{x}{2^{n\ell}}\right) + \left(\frac{1+\ell}{2}\right)|2|^{4n\ell}\tilde{\varphi}\left(\frac{x}{2^{n\ell}}\right)\right\}\end{aligned} \tag{3.181}$$

for all $x, y \in G$, and the limit

$$\begin{aligned}\lim_{n\to\infty}\max&\left\{\left[\left(\frac{1-\ell}{2}\right)|2|^{2\ell(j+\frac{1+\ell}{2})} + \left(\frac{1+\ell}{2}\right)|2|^{4\ell(j+\frac{1+\ell}{2})}\right]\right.\\ &\left.\cdot\,\tilde{\varphi}\left(\frac{x}{2^{\ell(j+\frac{1+\ell}{2})}}\right) : 0 \le j < n\right\}\end{aligned} \tag{3.182}$$

exists for all $x \in G$, where $\tilde{\varphi}(x)$ is defined as in Eq. [3.160] for all $x \in G$. Suppose that an even function $f : G \to X$ with $f(0) = 0$ satisfies the inequality [3.119] *for all $x, y \in G$. Then there exist a quadratic function $Q : G \to X$ and a quartic function $V : G \to X$ such that*

$$\|f(x) - Q(x) - V(x)\| \le \frac{1}{|48|}\max\left\{\psi_q(x), \frac{1}{|4|}\psi_v(x)\right\} \tag{3.183}$$

for all $x \in G$, where $\psi_q(x)$ and $\psi_v(x)$ are defined as in Theorems 3.3.8 and 3.3.9. Moreover, if

$$\begin{aligned}\lim_{i\to\infty}\lim_{n\to\infty}\max&\left\{\left[\left(\frac{1-\ell}{2}\right)|2|^{2\ell\left(j+\frac{1+\ell}{2}\right)} + \left(\frac{1+\ell}{2}\right)|2|^{4\ell\left(j+\frac{1+\ell}{2}\right)}\right]\right.\\ &\left.\cdot\,\tilde{\varphi}\left(\frac{x}{2^{\ell\left(j+\frac{1+\ell}{2}\right)}}\right) : i \le j < n+i\right\} = 0,\end{aligned}$$

then Q is the unique quadratic function and V is the unique quartic function satisfying Eq. [3.183].

Proof. The proof is similar to the proof of Theorem 3.3.6 and the result follows from Theorems 3.3.8 and 3.3.9. □

Theorem 3.3.11 (Eshaghi et al. [37]). *Let $\varphi : G \times G \to [0, \infty)$ be a function such that*

$$\lim_{n\to\infty} |2|^{2n}\varphi\left(\frac{x}{2^n}, \frac{y}{2^n}\right) = 0 = \lim_{n\to\infty} |2|^{2n}\tilde{\varphi}\left(\frac{x}{2^n}\right),$$
$$\lim_{n\to\infty} \frac{1}{|2|^{4n}}\varphi(2^n x, 2^n y) = 0 = \lim_{n\to\infty} \frac{1}{|2|^{4n}}\tilde{\varphi}(2^n x) \tag{3.184}$$

for all $x, y \in G$, and the limits

$$\lim_{n\to\infty} \max\left\{|2|^{2(j+1)}\tilde{\varphi}\left(\frac{x}{2^{j+1}}\right) : 0 \le j < n\right\}, \quad \lim_{n\to\infty} \max\left\{\frac{1}{|2|^{4j}}\tilde{\varphi}(2^j x) : 0 \le j < n\right\}$$

exist for all $x \in G$, where $\tilde{\varphi}(x)$ is defined as in Eq. [3.118] for all $x \in G$. Suppose that an even function $f : G \to X$ with $f(0) = 0$ satisfies the inequality [3.119] *for all $x, y \in G$. Then there exist a quadratic function $Q : G \to X$ and a quartic function $V : G \to X$ such that*

$$\|f(x) - Q(x) - V(x)\| \le \frac{1}{|48|} \max\left\{\psi_q(x), \frac{1}{|4|}\psi_v(x)\right\} \tag{3.185}$$

for all $x \in G$, where $\psi_q(x)$ and $\psi_v(x)$ are defined as in Theorems 3.3.8 and 3.3.9. Moreover, if

$$\lim_{i\to\infty}\lim_{n\to\infty} \max\left\{\frac{1}{|2|^{4j}}\tilde{\varphi}(2^j x) : i \le j < n + i\right\}$$
$$= 0 = \lim_{i\to\infty}\lim_{n\to\infty} \max\left\{|2|^{2(j+1)}\tilde{\varphi}\left(\frac{x}{2^{j+1}}\right) : 0 \le j < n\right\},$$

then Q is the unique quadratic function and V is the unique quartic function satisfying Eq. [3.185].

Now, we are ready to prove the main theorem concerning the generalized Hyers-Ulam stability problem for the Eq. [3.115] in non-Archimedean spaces.

Theorem 3.3.12 (Eshaghi et al. [37]). *Let $\ell \in \{1, -1\}$ be fixed and let $\varphi : G \times G \to [0, \infty)$ be a function satisfies Eqs. [3.152] and* [3.181] *for all $x, y \in G$, and*

$$\lim_{n\to\infty} \max\left\{\left[\left(\frac{1-\ell}{2}\right)|2|^{\ell(j+\frac{1+\ell}{2})} + \left(\frac{1+\ell}{2}\right)|2|^{3\ell(j+\frac{1+\ell}{2})}\right]\right.$$
$$\left.\cdot\,\tilde{\psi}\left(\frac{x}{2^{\ell(j+\frac{1+\ell}{2})}}\right) : 0 \le j < n\right\},$$

$$\lim_{n\to\infty} \max\left\{\left[\left(\frac{1-\ell}{2}\right)|2|^{2\ell(j+\frac{1+\ell}{2})} + \left(\frac{1+\ell}{2}\right)|2|^{4\ell(j+\frac{1+\ell}{2})}\right]\right.$$
$$\left.\cdot\,\tilde{\varphi}\left(\frac{x}{2^{\ell(j+\frac{1+\ell}{2})}}\right) : 0 \le j < n\right\}$$

exists for all $x \in G$, *where* $\tilde{\psi}(x)$ *and* $\tilde{\varphi}(x)$ *are defined as in Eqs. [3.118] and* [3.160] *for all* $x \in G$. *Suppose that a function* $f : G \to X$ *with* $f(0) = 0$ *satisfies the inequality* [3.119] *for all* $x, y \in G$. *Then there exist an additive function* $A : G \to X$, *a quadratic function* $Q : G \to X$, *a cubic function* $C : G \to X$ *and a quartic function* $V : G \to X$ *such that*

$$\begin{aligned} &\|f(x) - A(x) - Q(x) - C(x) - V(x)\| \\ &\leq \frac{1}{|24|} \max\left\{ \frac{1}{|4|} \max\left\{ \max\{\psi_q(x), \frac{1}{|4|}\psi_v(x)\}, \right.\right. \\ &\left. \max\{\psi_q(-x), \frac{1}{|4|}\psi_v(-x)\} \right\}, \max\left\{ \max\{\psi_a(x), \frac{1}{|4|}\psi_c(x)\}, \right. \\ &\left.\left. \max\{\psi_a(-x), \frac{1}{|4|}\psi_c(-x)\} \right\} \right\} \end{aligned} \tag{3.186}$$

for all $x \in G$, *where* $\psi_a(x)$, $\psi_c(x)$, $\psi_q(x)$ *and* $\psi_v(x)$ *are defined as in Theorems 3.3.4, 3.3.5, 3.3.8, and 3.3.9. Moreover, if*

$$\begin{aligned} \lim_{i\to\infty} \lim_{n\to\infty} \max &\left\{ \left[\left(\frac{1-\ell}{2} \right) |2|^{\ell(j+\frac{1+\ell}{2})} + \left(\frac{1+\ell}{2} \right) |2|^{3\ell(j+\frac{1+\ell}{2})} \right] \right. \\ &\left. \cdot \tilde{\psi}\left(\frac{x}{2^{\ell(j+\frac{1+\ell}{2})}} \right) : i \leq j < n+i \right\} = 0 \end{aligned}$$

$$\begin{aligned} \lim_{i\to\infty} \lim_{n\to\infty} \max &\left\{ \left[\left(\frac{1-\ell}{2} \right) |2|^{2\ell(j+\frac{1+\ell}{2})} + \left(\frac{1+\ell}{2} \right) |2|^{4\ell(j+\frac{1+\ell}{2})} \right] \right. \\ &\left. \cdot \tilde{\varphi}\left(\frac{x}{2^{\ell(j+\frac{1+\ell}{2})}} \right) : i \leq j < n+i \right\} = 0, \end{aligned}$$

then A *is the unique additive function,* Q *is the unique quadratic function,* C *is the unique cubic function, and* V *is the unique quartic function satisfying Eq. [3.186].*

Proof. Let $\ell = 1$ and $f_o(x) = \frac{1}{2}(f(x) - f(-x))$ for all $x \in G$. Then

$$\|Df_o(x, y)\| \leq \frac{1}{|2|} \max\{\varphi(x, y), \varphi(-x, -y)\}$$

for all $x, y \in G$. From Theorem 3.3.6, it follows that there exist a unique additive function $A : G \to X$ and a unique cubic function $C : G \to X$ satisfying

$$\begin{aligned} &\|f_o(x) - A(x) - C(x)\| \\ &\leq \frac{1}{|24|} \max\left\{ \max\left\{ \psi_a(x), \frac{1}{|4|}\psi_c(x) \right\}, \max\left\{ \psi_a(-x), \frac{1}{|4|}\psi_c(-x) \right\} \right\} \end{aligned} \tag{3.187}$$

for all $x \in G$. Also, let $f_e(x) = \frac{1}{2}(f(x) + f(-x))$ for all $x \in G$. Then

$$\|Df_e(x, y)\| \leq \frac{1}{|2|} \max\{\varphi(x, y), \varphi(-x, -y)\}$$

for all $x, y \in G$. From Theorem 3.3.10, it follows that there exist a quadratic function $Q : G \to X$ and a quartic function $V : G \to X$ satisfying

$$\|f_e(x) - Q(x) - V(x)\| \leq \frac{1}{|96|} \max\left\{\max\left\{\psi_q(x), \frac{1}{|4|}\psi_v(x)\right\}, \max\left\{\psi_q(-x), \frac{1}{|4|}\psi_v(-x)\right\}\right\} \tag{3.188}$$

for all $x \in G$. Hence, Eq. [3.186] follows from Eqs. [3.187] and [3.188]. The rest of the proof is trivial. For $\ell = -1$, we can prove the theorem by a similar technique. □

Theorem 3.3.13 ([37]). *Let $\varphi : G \times G \to [0, \infty)$ be a function satisfies Eqs. [3.156] and* [3.184] *for all $x, y \in G$, and the limits*

$$\lim_{n\to\infty} \max\left\{|2|^{j+1}\tilde{\psi}\left(\frac{x}{2^{j+1}}\right) : 0 \leq j < n\right\}, \quad \lim_{n\to\infty} \max\left\{\frac{1}{|2|^{3j}}\tilde{\psi}(2^j x) : 0 \leq j < n\right\},$$
$$\lim_{n\to\infty} \max\left\{|2|^{2(j+1)}\tilde{\varphi}\left(\frac{x}{2^{j+1}}\right) : 0 \leq j < n\right\}, \quad \lim_{n\to\infty} \max\left\{\frac{1}{|2|^{4j}}\tilde{\varphi}(2^j x) : 0 \leq j < n\right\}$$

exist for all $x \in G$, where $\tilde{\psi}(x)$ and $\tilde{\varphi}(x)$ are defined as in Eqs. [3.118] and [3.160] *for all $x \in G$. Suppose that a function $f : G \to X$ with $f(0) = 0$ satisfies the inequality* [3.119] *for all $x, y \in G$. Then there exist an additive function $A : G \to X$, a quadratic function $Q : G \to X$, a cubic function $C : G \to X$ and a quartic function $V : G \to X$ such that*

$$\begin{aligned} &\|f(x) - A(x) - Q(x) - C(x) - V(x)\| \\ &\leq \frac{1}{|24|} \max\left\{\frac{1}{|4|} \max\left\{\max\left\{\psi_q(x), \frac{1}{|4|}\psi_v(x)\right\},\right.\right. \\ &\left.\left.\max\left\{\psi_q(-x), \frac{1}{|4|}\psi_v(-x)\right\}\right\}, \max\left\{\max\left\{\psi_a(x), \frac{1}{|4|}\psi_c(x)\right\},\right.\right. \\ &\left.\left.\max\left\{\psi_a(-x), \frac{1}{|4|}\psi_c(-x)\right\}\right\}\right\} \end{aligned} \tag{3.189}$$

for all $x \in G$, where $\psi_a(x)$, $\psi_q(x)$, $\psi_c(x)$, and $\psi_v(x)$ are defined as in Theorems 3.3.4, 3.3.5, 3.3.8, and 3.3.9. Moreover, if

$$\begin{aligned} &\lim_{i\to\infty}\lim_{n\to\infty} \max\left\{|2|^{j+1}\tilde{\psi}\left(\frac{x}{2^{j+1}}\right) : i \leq j < n+i\right\} \\ &= \lim_{i\to\infty}\lim_{n\to\infty} \max\left\{\frac{1}{|2|^{3j}}\tilde{\psi}(2^j x) : i \leq j < n+i\right\} \\ &= 0 = \lim_{i\to\infty}\lim_{n\to\infty} \max\left\{\frac{1}{|2|^{4j}}\tilde{\varphi}(2^j x) : i \leq j < n+i\right\} \\ &= \lim_{i\to\infty}\lim_{n\to\infty} \max\left\{|2|^{2(j+1)}\tilde{\varphi}\left(\frac{x}{2^{j+1}}\right) : 0 \leq j < n\right\}, \end{aligned}$$

then A is the unique additive function, Q is the unique quadratic function, C is the unique cubic function, and V is the unique quartic function satisfying Eq. [3.189].

Proof. The proof is similar to the proof of Theorem 3.3.12 and the result follows from Theorems 3.3.7 and 3.3.11. □

Stability of functional equations in Banach algebras

4

4.1 Approximate homomorphisms and derivations in ordinary Banach algebras

Let $\mathcal{A}$ and $\mathcal{B}$ be two Banach algebras. A function $f : \mathcal{A} \to \mathcal{B}$ is called a ring homomorphism or additive homomorphism if f is an additive function satisfying

$$f(xy) = f(x)f(y)$$

for all $x, y \in \mathcal{A}$. Bourgin [38, 39] is the first mathematician dealing with the stability of (ring) homomorphisms in Banach algebras. Since then, the topic of approximate homomorphisms has been studied by a number of mathematicians: see [7, 40, 41] and references therein.

Let $\mathcal{A}_1$ be a subalgebra of Banach algebra $\mathcal{A}$. A function $f : \mathcal{A}_1 \to \mathcal{A}$ is called a derivation if and only if it satisfies the functional equations

$$\begin{aligned} d(x+y) &= d(x) + d(y), \\ d(xy) &= xd(y) + d(x)y \end{aligned}$$

for all $x, y \in \mathcal{A}_1$.

Theorem 4.1.1 (Bourgin [38]). *Let ε and δ be nonnegative real numbers. Then every mapping f of a Banach algebra $\mathcal{A}$ with an identity element onto a Banach algebra $\mathcal{B}$ with an identity element satisfying*

$$\|f(x+y) - f(x) - f(y)\| \leq \varepsilon \tag{4.1}$$

and

$$\|f(x \cdot y) - f(x)f(y)\| \leq \delta \tag{4.2}$$

for all $x, y \in \mathcal{A}$, is a ring homomorphism of $\mathcal{A}$ onto $\mathcal{B}$.

Badora [40] gave a simple proof of the following generalization of the Bourgin's result.

Theory of Approximate Functional Equations. http://dx.doi.org/10.1016/B978-0-12-803920-5.00004-3

Theorem 4.1.2 (Badora [40]). *Let $\mathcal{R}$ be a ring, let $\mathcal{B}$ be a Banach algebra and let ε and δ be nonnegative real numbers. Assume that $f : \mathcal{R} \to \mathcal{B}$ satisfies Eqs. [4.1] and* [4.2] *for all $x, y \in \mathcal{R}$. Then there exists a unique ring homomorphism $h : \mathcal{R} \to \mathcal{B}$ such that*

$$\|f(x) - h(x)\| \leq \varepsilon \tag{4.3}$$

for all $x \in \mathcal{R}$. Moreover,

$$b \cdot \big(f(x) - h(x)\big) = 0, \quad \big(f(x) - h(x)\big) \cdot b = 0, \tag{4.4}$$

for all $x \in \mathcal{R}$ and all b from the algebra generated by $h(\mathcal{R})$.

Proof. The Hyers theorem (Theorem 2.1.1) shows that there exists an additive function $h : \mathcal{R} \to \mathcal{B}$ such that

$$\|f(x) - h(x)\| \leq \varepsilon$$

for all $x \in \mathcal{R}$. Now we only need to show that h is a multiplicative function. Our inequality follows that

$$\|f(nx) - h(nx)\| \leq \varepsilon$$

for all $x \in \mathcal{R}$ and $n \in \mathbb{N}$. By the additivity of h it is easy to see that then

$$\left\| \frac{1}{n} f(nx) - h(nx) \right\| \leq \frac{1}{n} \varepsilon$$

for all $x \in \mathcal{R}$ and $n \in \mathbb{N}$, which means that

$$h(x) = \lim_{n \to \infty} \frac{1}{n} f(nx) \tag{4.5}$$

for all $x \in \mathcal{R}$. Let

$$r(x, y) = f(x \cdot y) - f(x) f(y)$$

for all $x, y \in \mathcal{R}$. Then, using inequality [4.2], we get

$$\lim_{n \to \infty} r(nx, y) = 0 \tag{4.6}$$

for all $x, y \in \mathcal{R}$. Applying Eqs. [4.5] and [4.6], we have

$$\begin{aligned} h(x \cdot y) &= \lim_{n \to \infty} \frac{1}{n} f\big(n(x \cdot y)\big) = \lim_{n \to \infty} \frac{1}{n} f\big((nx) \cdot y\big) \\ &= \lim_{n \to \infty} \frac{1}{n} \big(f(nx) f(y) + r(nx, y)\big) = h(x) f(y) \end{aligned}$$

for all $x, y \in \mathcal{R}$. The result of our calculation is the following functional equation

$$h(x \cdot y) = h(x)f(y) \tag{4.7}$$

for all $x, y \in \mathcal{R}$. From this equation by the additivity of h, we have

$$h(x)f(ny) = h\big(x \cdot (ny)\big) = h\big((nx) \cdot y\big) = h(nx)f(y) = nh(x)f(y)$$

for all $x, y \in \mathcal{R}$ and $n \in \mathbb{N}$. Therefore,

$$h(x)\frac{1}{n}f(ny) = h(x)f(y)$$

for all $x, y \in \mathcal{R}$ and $n \in \mathbb{N}$. Sending n to infinity, by Eq. [4.5], we see that

$$h(x)h(y) = h(x)f(y) \tag{4.8}$$

for all $x, y \in \mathcal{R}$. Combining this formula with Eq. [4.7] we have that h is a multiplicative function which is the desired conclusion.

To prove the uniqueness property of h, assume that h^* is another ring homomorphism with

$$\|f(x) - h^*(x)\| \leq \varepsilon$$

for all $x \in \mathcal{R}$. Since both h and h^* are additive, we deduce that

$$n\|h(x) - h^*(x)\| = \|h(nx) - h^*(nx)\| \leq 2\varepsilon,$$

so that

$$\|h(x) - h^*(x)\| \leq \frac{2\varepsilon}{n}$$

for all $x \in \mathcal{R}$ and $n \in \mathbb{N}$. Letting n to infinity we find that $h(x) = h^*(x)$ for all $x \in \mathcal{R}$. Moreover, from Eq. [4.2] we get

$$\lim_{n\to\infty} \frac{1}{n}r(x, ny) = 0$$

for all $x, y \in \mathcal{R}$. Thus, by Eq. [4.5], we deduce that

$$\begin{aligned} h(x \cdot y) &= \lim_{n\to\infty} \frac{1}{n}f\big(n(x \cdot y)\big) = \lim_{n\to\infty} \frac{1}{n}f\big(x \cdot (ny)\big) \\ &= \lim_{n\to\infty} \left(f(x)\frac{1}{n}f(ny) + \frac{1}{n}r(x, ny)\right) = f(x)h(y) \end{aligned}$$

for all $x, y \in \mathcal{R}$. Hence, with regard to Eq. [4.8],

$$f(x)h(y) = h(x \cdot y) = h(x)h(y) = h(x)f(y)$$

for all $x, y \in \mathcal{R}$. This identity leads to

$$h(x) \cdot \big(f(y)h(y)\big) = 0, \quad \big(f(y)h(y)\big) \cdot h(x) = 0, \tag{4.9}$$

for all $x, y \in \mathcal{R}$, which shows Eq. [4.8] and completes the proof of Theorem 4.1.1. □

In particular, using $\varepsilon(\|x\|^p + \|y\|^p)$ and $\delta(\|x\|^q\|y\|^q)$ instead of ε and δ, respectively, for $\varepsilon, \delta \geq 0$ and some real numbers p, q in the main theorem, one gets the following corollary as a consequence of Rassias theorem (Theorem 2.2.1).

Corollary 4.1.3 (Badora [40]). *Let $\mathcal{A}$ be a normed algebra and let $\mathcal{B}$ be a Banach algebra. Moreover, let ε and δ be nonnegative real numbers and let p and q be a real numbers such that $p, q < 1$ or $p, q > 1$. Assume that $f : \mathcal{A} \to \mathcal{B}$ satisfies the system of functional inequalities*

$$\|f(x + y) - f(x) - f(y)\| \leq \varepsilon(\|x\|^p + \|y\|^p) \tag{4.10}$$

and

$$\|f(x \cdot y) - f(x)f(y)\| \leq \delta(\|x\|^q\|y\|^q) \tag{4.11}$$

for all $x, y \in \mathcal{A}$. Then there exists a unique ring homomorphism $h : \mathcal{A} \to \mathcal{B}$ and a constant k such that

$$\|f(x) - h(x)\| \leq K\varepsilon\|x\|^p \tag{4.12}$$

for all $x \in \mathcal{A}$.

Remark 4.1.4 (Badora [40]). Under the hypothesis of Corollary 4.1.3 for real Banach algebras, if we assume that, for each fixed $x \in \mathcal{A}$ and for all $t \in \mathbb{R}$, the function $t \mapsto f(tx)$ is continuous then, by the Rassias theorem, the additive mapping h is linear, so h is an algebra homomorphism.

Badora also proved the stability (in the sense of Hyers) of derivations in Banach algebras.

Theorem 4.1.5 (Badora [42]). *Let $\mathcal{A}_1$ be a closed subalgebra of a Banach algebra $\mathcal{A}$. Assume that $f : \mathcal{A}_1 \to \mathcal{A}$ satisfies*

$$\|f(x + y) - f(x) - f(y)\| \leq \delta \quad (x, y \in \mathcal{A}_1)$$

and

$$\|f(xy) - xf(x) - f(x)y)\| \leq \varepsilon \quad (x, y \in \mathcal{A}_1)$$

for some constants $\delta, \varepsilon \geq 0$. Then there exists a unique derivation $d : \mathcal{A}_1 \to \mathcal{A}$ such that

$$\|f(x) - d(x)\| \leq \delta \quad (x \in \mathcal{A}_1).$$

Furthermore,

$$x(f(y) - h(y)) = (f(y) - h(y))x = 0 \tag{4.13}$$

for all $x, y \in \mathcal{A}_1$.

Let $\mathcal{X}$ be a Banach bimodule over a Banach algebra $\mathcal{A}$. Recall that $\mathcal{X} \oplus_1 \mathcal{A}$ is a Banach algebra equipped with the following ℓ_1-norm

$$\|(x, a)\| = \|x\| + \|a\| \quad (a \in \mathcal{A}, x \in \mathcal{X})$$

and the product

$$(x_1, a_1)(x_2, a_2) = (x_1 \cdot a_2 + a_1 \cdot x_2, a_1 a_2) \quad (a_1, a_2 \in \mathcal{A}, x_1, x_2 \in \mathcal{X}).$$

The algebra $\mathcal{X} \oplus_1 \mathcal{A}$ is called a module extension Banach algebra. We also define (norm decreasing) projection maps $\pi_1 : \mathcal{X} \oplus_1 \mathcal{A} \to \mathcal{X}$ and $\pi_2 : \mathcal{X} \oplus_1 \mathcal{A} \to \mathcal{A}$ by $(x, b) \mapsto x$ and $(x, b) \mapsto b$, respectively. We refer the reader to [43] for more information on Banach modules and to [44] for details on module extensions. The following theorem may be regarded as an extension of Theorem 4.1.5.

Theorem 4.1.6. *Let $\varepsilon, \delta > 0$. Let $\mathcal{A}$ be a Banach algebra and let $\mathcal{X}$ be a Banach A-bimodule. Suppose that a function $f : \mathcal{A} \to \mathcal{X}$ satisfies*

$$\|f(a + b) - f(a) - f(b)\| \leq \varepsilon \tag{4.14}$$

and

$$\|f(ab) - f(a)b - af(b)\| \leq \delta$$

for all $a, b \in \mathcal{A}$. Then there exists a unique derivation $D : A \to \mathcal{X}$ such that

$$\|f(a) - D(a)\| \leq \varepsilon \quad (a \in \mathcal{A}).$$

Moreover, we have

$$b(f(a) - D(a)) = (f(a) - D(a))b = 0 \quad (a, b \in \mathcal{A}). \tag{4.15}$$

Proof. Let us define the mapping $\varphi_f : \mathcal{A} \to \mathcal{X} \oplus_1 \mathcal{A}$ by $a \mapsto (f(a), a)$. We have

$$\begin{aligned}
&\|\varphi_f(a + b) - \varphi_f(a) - \varphi_f(b)\| \\
&\quad = \|(f(a + b), a + b) - (f(a), a) - (f(b), b)\|
\end{aligned}$$

$$= \|(f(a+b) - f(a) - f(b), 0)\|$$
$$= \|f(a+b) - f(a) - f(b)\| \leq \varepsilon$$

and similarly

$$\|\varphi_f(ab) - \varphi_f(a)\varphi_f(b)\| \leq \delta$$

for all $a, b \in \mathcal{A}$. It follows from Theorem 4.1.1 that there exists a unique ring homomorphism $\varphi : \mathcal{A} \to \mathcal{X} \oplus_1 \mathcal{A}$ such that

$$\|\varphi(a) - \varphi_f(a)\| \leq \varepsilon \quad (a \in \mathcal{A}). \tag{4.16}$$

Moreover,

$$(x, b)(\varphi(a) - \varphi_f(a)) = (\varphi(a) - \varphi_f(a))(x, b) = 0 \tag{4.17}$$

for all $a \in \mathcal{A}$, and all (x, b) in the algebra generated by $\varphi(\mathcal{A})$.

It follows from Eq. [4.16] that

$$\|(\pi_2 \circ \varphi_f)(na) - (\pi_2 \circ \varphi)(na)\| \leq \|\varphi_f(na) - \varphi(na)\| \leq \varepsilon \tag{4.18}$$

for all $n \in \mathbb{N}, a \in \mathcal{A}$. By the additivity of mappings under consideration

$$(\pi_2 \circ \varphi)(na) = n(\pi_2 \circ \varphi)(a)$$

and

$$(\pi_2 \circ \varphi_f)(na) = \pi_2(f(na), na) = na,$$

whence, by Eq. [4.18],

$$\|a - (\pi_2 \circ \varphi)(a)\| \leq \frac{1}{n}\varepsilon \tag{4.19}$$

for all $n \in \mathbb{N}, a \in \mathcal{A}$. By letting n tend to ∞ in Eq. [4.19], we obtain

$$(\pi_2 \circ \varphi)(a) = a \quad (a \in \mathcal{A}). \tag{4.20}$$

Hence,

$$\begin{aligned}
\big((\pi_1 \circ \varphi)(ab), ab\big) &= \big(\pi_1(\varphi(ab)), \pi_2(\varphi(ab))\big) = \varphi(ab) = \varphi(a)\varphi(b) \\
&= \big(\pi_1(\varphi(a)), \pi_2(\varphi(a))\big)\big(\pi_1(\varphi(b)), \pi_2(\varphi(b))\big) \\
&= \big(\pi_1(\varphi(a)), a\big)\big(\pi_1(\varphi(b)), b\big) \\
&= \big(a\pi_1(\varphi(b)) + \pi_1(\varphi(a))b, ab\big)
\end{aligned} \tag{4.21}$$

for all $a, b \in \mathcal{A}$. Put $D := \pi_1 \circ \varphi$. Then it follows from Eq. [4.21] that D is a derivation from $\mathcal{A}$ into $\mathcal{X}$. It follows from Eq. [4.16] that

$$\|D(a) - f(a)\| = \|\pi_1(\varphi(a)) - \pi_1(\varphi_f(a))\| \leq \|\varphi(a) - \varphi_f(a)\| \leq \varepsilon$$

for all $a \in \mathcal{A}$.

To prove the uniqueness of D, assume that D^* is another derivation from $\mathcal{A}$ into $\mathcal{X}$ satisfying

$$\|D^*(a) - f(a)\| \leq \varepsilon \quad (a \in \mathcal{A}).$$

Then

$$\begin{aligned}\|D(a) - D^*(a)\| &= \frac{1}{n}\|D(na) - D^*(na)\| \\ &\leq \frac{1}{n}\|D^*(a) - f(a)\| + \frac{1}{n}\|D(a) - f(a)\| \\ &\leq \frac{2}{n}\varepsilon\end{aligned}$$

for all $a \in \mathcal{A}, n \in \mathbb{N}$. By letting $n \to \infty$ in the last inequality, we conclude that $D(a) = D^*(a)$ for all $a \in \mathcal{A}$.

Moreover, by Eq. [4.17],

$$\begin{aligned}(f(a) - D(a))b &= \pi_1\big((f(a) - D(a))b, 0\big) \\ &= \pi_1\Big(\big(f(a) - D(a), 0\big)\big(D(b), b\big)\Big) \\ &= \pi_1\Big(\big(\pi_1(\varphi(a) - \varphi_f(a)), 0\big)\big(\pi_1(\varphi(b)), b\big)\Big) \\ &= \pi_1\Big(\big(\pi_1(\varphi(a) - \varphi_f(a)), 0\big)\varphi(b)\Big) \\ &= \pi_1\Big(\big((\pi_1(\varphi(a)), a\big) - \big(\pi_1(\varphi_f(a)), a\big)\big)\varphi(b)\Big) \\ &= \pi_1\Big(\big(\varphi(a) - \varphi_f(a)\big)\varphi(b)\Big) \\ &= \pi_1(0, 0) \\ &= 0\end{aligned}$$

for all $a, b \in \mathcal{A}$. Similarly we have $b(f(a) - D(a)) = 0$ for all $a, b \in \mathcal{A}$. □

Remark 4.1.7. To achieve an algebra homomorphism, ie, a homogeneous ring homomorphism, one can replace inequality [4.14] by

$$\|f(\lambda a + b) - \lambda f(a) - f(b)\| \leq \varepsilon$$

where $\lambda \in \{z \in \mathbb{C} : |z| = 1\}$ and $a, b \in \mathcal{A}$. Then a standard argument shows that D turns into a linear derivation.

We now present two superstability results concerning derivations.

Corollary 4.1.8. *Let $\varepsilon, \delta > 0$. Let $\mathcal{A}$ be a Banach algebra and let $\mathcal{X}$ be a Banach $\mathcal{A}$-bimodule without order, ie, $\mathcal{A}x = 0$ or $x\mathcal{A} = 0$ implies that $x = 0$, where $x \in \mathcal{X}$. Suppose that a function $f : \mathcal{A} \to \mathcal{X}$ satisfies*

$$\|f(a+b) - f(a) - f(b)\| \leq \varepsilon$$

and

$$\|f(ab) - f(a)b - af(b)\| \leq \delta$$

for all $a, b \in \mathcal{A}$. Then f is a derivation.

Proof. Due to $\mathcal{X}$ is a Banach $\mathcal{A}$-bimodule without order, Eq. [4.15] in Theorem 4.1.5 implies that $f = d$ is a derivation. □

Corollary 4.1.9. *Let $\varepsilon, \delta > 0$. Let $\mathcal{A}$ be a Banach algebra with a bounded approximate identity. Suppose that a function $f : \mathcal{A} \to \mathcal{A}$ satisfies*

$$\|f(a+b) - f(a) - f(b)\| \leq \varepsilon$$

and

$$\|f(ab) - f(a)b - af(b)\| \leq \delta$$

for all $a, b \in \mathcal{A}$. Then f is a derivation.

Proof. This follows from Corollary 4.1.8, since every Banach algebra with an approximate unit, as a Banach bimodule over itself, is without order. □

Similarly, we can use Theorem 4.1.1 to prove the Hyers-Ulam stability of derivations as follows.

Theorem 4.1.10. *Let $\varepsilon, \delta > 0$, and let p, q be real numbers such that $p, q < 1$, or $p, q > 1$. Let $\mathcal{A}$ be a Banach algebra and let $\mathcal{X}$ be a Banach A-bimodule. Suppose that a function $f : \mathcal{A} \to \mathcal{X}$ satisfies*

$$\|f(a+b) - f(a) - f(b)\| \leq \varepsilon(\|a\|^p + \|b\|^p)$$

and

$$\|f(ab) - f(a)b - af(b)\| \leq \delta(\|a\|^q \|b\|^q)$$

for all $a, b \in \mathcal{A}$. Then there exists a unique derivation $D : A \to \mathcal{X}$ and a constant k such that

$$\|f(a) - D(a)\| \leq k\varepsilon \|x\|^p$$

for all $a \in \mathcal{A}$.

The following counterexample, which is a modification of Luminet's example (see [45]), shows that this result is failed for $p = 1$.

Example 4.1.11. Let

$$X = \begin{bmatrix} 0 & 0 & 0 \\ \mathbb{R} & 0 & 0 \\ 0 & 0 & 0 \end{bmatrix}$$

and define a function $\varphi : \mathbb{R} \to \mathbb{R}$ by

$$\varphi(x) = \begin{cases} 0 & |x| \le 1, \\ x \ln(|x|) & |x| > 1. \end{cases}$$

Let $f : X \to X$ be defined by

$$f\left(\begin{bmatrix} 0 & 0 & 0 \\ x & 0 & 0 \\ 0 & 0 & 0 \end{bmatrix}\right) = \begin{bmatrix} 0 & 0 & 0 \\ \varphi(x) & 0 & 0 \\ 0 & 0 & 0 \end{bmatrix}$$

for all $x \in \mathbb{R}$. Then

$$\|f(a+b) - f(a) - f(b)\| \le \varepsilon(\|a\| + \|b\|)$$

and

$$\|f(ab) - f(a)b - af(b)\| \le \delta(\|a\|^2 \|b\|^2)$$

for some $\delta > 0, \varepsilon > 0$ and all $a, b \in X$; see [10]. Therefore, f satisfies the conditions of Theorem 4.1.10 with $p = 1$, $q = 2$. There is, however, no derivation $D : X \to X$ and no constant $k > 0$ such that

$$\|f(a) - D(a)\| \le k\varepsilon \|a\| \quad (a \in X).$$

Against this, assume that there exist a derivation $D : X \to X$ and a constant $k > 0$ such that

$$\|f(a) - D(a)\| \le k\varepsilon \|a\| \quad (a \in X).$$

Representing D as a 3×3 matrix $[D_{ij}]$, we infer that $D_{21} : \mathbb{R} \to \mathbb{R}$ is an additive mapping such that

$$|\varphi(x) - D_{21}(x)| \le k\varepsilon |x|$$

for all $x \in \mathbb{R}$. From the continuity of φ, it follows that D_{21} is bounded on some neighborhood of zero. Then there exists a fixed c such that $D_{21}(x) = cx$, for all $x \in \mathbb{R}$. Hence,

$$|x\ln(x) - cx| \leq k\varepsilon x$$

for $x > 1$, whence

$$|\ln(x) - c| \leq k\varepsilon$$

for $x > 1$, which yields a contradiction.

4.2 Approximate homomorphisms and derivations in C^*-algebras

A C^*-algebra $\mathcal{A}$ endowed with the Lie product $[x, y] = xy - yx$ on $\mathcal{A}$ is called a Lie C^*-algebra.

Definition 4.2.1. A $\mathbb{C}$-linear and $*$-preserving mapping H of a Lie C^*-algebra $\mathcal{A}$ to a Lie C^*-algebra $\mathcal{B}$ is called Lie $*$-homomorphism if

$$H\big([x, y]\big) = [H(x), H(y)] \tag{4.22}$$

holds for all $x, y \in \mathcal{A}$.

Definition 4.2.2. Let $\mathcal{A}$ be a Lie C^*-algebra and $\mathcal{A}_1$ be a Lie C^*-subalgebra of $\mathcal{A}$. A $\mathbb{C}$-linear and $*$-preserving mapping $D : \mathcal{A}_1 \to \mathcal{A}$ is called Lie $*$-derivation if

$$D\big([x, y]\big) = [D(x), y] + [x, D(y)] \tag{4.23}$$

holds for all $x, y \in \mathcal{A}_1$.

Park [46] proved the generalized Hyers-Ulam stability of Lie $*$-homomorphisms in Lie C^*-algebras, and of Lie $*$-derivations on Lie C^*-algebras.

Let $\mathcal{A}$ be a Lie C^*-algebra with unit e and $\mathcal{B}$ be Lie C^*-algebra. Let $\mathcal{U}(\mathcal{A}) = \{u \in \mathcal{A} | uu^* = u^*u = e\}$ and $\mathbb{T}^1 = \{\lambda \in \mathbb{C} : |\lambda| = 1\}$.

Theorem 4.2.3 (Park [46]). *Let $h : \mathcal{A} \to \mathcal{B}$ be a mapping with $h(0) = 0$ for which there exists a function $\varphi : \mathcal{A}^4 \to [0, \infty)$ such that*

$$\tilde{\varphi}(x, y, z, w) = \sum_{j=0}^{\infty} 3^{-j} \varphi\left(3^j x, 3^j y, 3^j z, 3^j w\right) < \infty,$$

$$\left\| 2h\left(\frac{\mu x + \mu y + [z, w]}{2}\right) - \mu h(x) - \mu h(y) - [h(z), h(w)] \right\| \leq \varphi(x, y, z, w),$$

$$\left\| h(3^n u^*) - h(3^n u)^* \right\| \leq \varphi(3^n u, 3^n u, 0, 0)$$

for all $\mu \in \mathbb{T}^1$, *all* $u \in \mathcal{U}(\mathcal{A})$, $n = 0, 1, \ldots$, *and all* $x, y, z, w \in \mathcal{A}\backslash\{0\}$. *Then there exists a unique Lie* $*$*-homomorphism* $L : \mathcal{A} \to \mathcal{B}$ *such that*

$$\|h(x) - L(x)\| \leq \frac{1}{3}\left(\tilde{\varphi}(x, -x, 0, 0) + \tilde{\varphi}(-x, 3x, 0, 0)\right)$$

for all $x \in \mathcal{A}\backslash\{0\}$.

Theorem 4.2.4 (Park [46]). *Let* $h : \mathcal{A} \to \mathcal{A}$ *be a mapping with* $h(0) = 0$ *for which there exists a function* $\varphi : \mathcal{A}^4 \to [0, \infty)$ *such that*

$$\tilde{\varphi}(x, y, z, w) = \sum_{j=0}^{\infty} 2^{-j} \varphi\left(2^j x, 2^j y, 2^j z, 2^j w\right) < \infty,$$

$$\|2h\left(\mu x + \mu y + [z, w]\right) - \mu h(x) - \mu h(y) - [h(z), w] - [z, h(w)]\| \leq \varphi(x, y, z, w),$$

$$\left\|h(2^n u^*) - h(2^n u)^*\right\| \leq \varphi(2^n u, 2^n u, 0, 0)$$

for all $\mu \in \mathbb{T}^1$, *all* $u \in \mathcal{U}(\mathcal{A})$, $n = 0, 1, \ldots$, *and all* $x, y, z, w \in \mathcal{A}$. *Then there exists a unique Lie* $*$*-derivation* $D : \mathcal{A} \to \mathcal{A}$ *such that*

$$\|h(x) - D(x)\| \frac{1}{2}\tilde{\varphi}(x, x, 0, 0)$$

for all $x \in \mathcal{A}$.

Park et al. [47] investigated $*$-homomorphisms between unital C^*-algebras.

Theorem 4.2.5 (Park et al. [47]). *Let* $h : \mathcal{A} \to \mathcal{B}$ *be a mapping satisfying* $h(0) = 0$ *and* $h(3^n uy) = h(3^n u)h(y)$ *for all* $u \in \mathcal{U}(\mathcal{A})$, $y \in \mathcal{A}$ *and all* $n = 0, 1, \ldots$, *for which there exists a function* $\varphi : \mathcal{A}\ \{0\} \times \mathcal{A}\ \{0\} \to [0, \infty)$ *such that*

$$\tilde{\varphi}(x, y) = \sum_{j=0}^{\infty} 3^{-j} \varphi\left(3^j x, 3^j y\right) < \infty,$$

$$\left\|2h\left(\frac{\mu x + \mu y}{2}\right) - \mu h(x) - \mu h(y)\right\| \leq \varphi(x, y),$$

$$\left\|h(3^n u^*) - h(3^n u)^*\right\| \leq \varphi(3^n u, 3^n u)$$

for all $\mu \in \mathbb{T}^1$, *all* $u \in \mathcal{U}(\mathcal{A})$, $n = 0, 1, \ldots$, *and all* $x, y \in \mathcal{A}$. *Assume that* $\lim_{n\to\infty} \frac{h(3^n e)}{3^n}$ *is invertible. Then the mapping* $h : \mathcal{A} \to \mathcal{B}$ *is a* $*$*-homomorphism.*

Eshaghi and Khodaei [48] dealt with the functional equation

$$\sum_{i=1}^{n} f\left(x_i + \frac{1}{n-1}\sum_{j=1, j\neq i}^{n} x_j\right) = 2\sum_{i=1}^{n} f(x_i), \tag{4.24}$$

where $n \in \mathbb{N}$ is a fixed integer with $n \geq 2$, and then they applied a fixed point theorem to investigate the stability by using contractively subhomogeneous and expansively superhomogeneous functions for Lie (α, β, γ)-derivations associated to the generalized Cauchy-Jensen functional equation [4.24] on Lie C^*-algebras. We observe that in case $n = 2$, Eq. [4.24] yields Cauchy-additive equation $f(x_1 + x_2) = f(x_1) + f(x_2)$.

Following [49], a $\mathbb{C}$-linear mapping $\mathfrak{D} : \mathcal{A} \to \mathcal{A}$ is a called a Lie (α, β, γ)-derivation of $\mathcal{A}$, if there exist $\alpha, \beta, \gamma \in \mathbb{C}$ such that

$$\alpha\mathfrak{D}[x, y] = \beta[\mathfrak{D}(x), y] + \gamma[x, \mathfrak{D}(y)]$$

for all $x, y \in \mathcal{A}$.

Let k be a fixed positive integer. We recall that a function $\rho : A \to B$, having a domain A and a codomain $(B, \leq)$ that are both closed under addition is called:

- a contractively subadditive function if there exists a constant L with $0 < L < 1$ such that $\rho(x + y) \leq L(\rho(x) + \rho(y))$;
- an expansively superadditive function if there exists a constant L with $0 < L < 1$ such that $\rho(x + y) \geq \frac{1}{L}(\rho(x) + \rho(y))$;
- a homogeneous function of degree k if $\rho(\lambda x) = \lambda^k \rho(x)$ (for the case of $k = 1$, the corresponding function is simply said to be homogeneous);
- a contractively subhomogeneous function of degree k if thereexists a constant L with $0 < L < 1$ such that $\rho(\lambda x) \leq L\lambda^k \rho(x)$;
- an expansively superhomogeneous function of degree k if there exists a constant L with $0 < L < 1$ such that $\rho(\lambda x) \geq \frac{\lambda^k}{L}\rho(x)$, for all $x, y \in A$ and all positive integer $\lambda > 1$.

Remark 4.2.6. If ρ is contractively subadditive and expansively superadditive separately, then ρ is contractively subhomogeneous $(:\ell = 1)$ and expansively superhomogeneous $(:\ell = -1)$, respectively, and so

$$\rho(\lambda^{\ell j} x) \leq \left(\lambda^{\ell} L\right)^j \rho(x), \quad j \in \mathbb{N}.$$

Also, if there exists a constant L with $0 < L < 1$ such that a function $\rho : A^n = \overbrace{A \times \cdots \times A}^{n\text{-times}} \to B$ satisfies

$$\rho\left(x_1, \ldots, \overbrace{\lambda^{\ell} x}^{i\text{th}}, \ldots, x_n\right) \leq \lambda^{\ell} L \rho\left(x_1, \ldots, \overbrace{x}^{i\text{th}}, \ldots, x_n\right)$$

for all $x, x_j \in A$ $(1 \leq j \neq i \leq n)$ and all positive integer λ, then we say that ρ is n-contractively subhomogeneous if $\ell = 1$, and ρ is n-expansively superhomogeneous if $\ell = -1$. It follows by the last inequality that ρ satisfies the properties

$$\rho\left(x_1,\ldots,\overbrace{\lambda^{\ell k}x}^{i\text{th}},\ldots,x_n\right) \le \left(\lambda^{\ell}L\right)^k \rho\left(x_1,\ldots,\overbrace{x}^{i\text{th}},\ldots,x_n\right), \quad k\in\mathbb{N},$$

$$\rho\left(\lambda^{\ell}x,\ldots,\lambda^{\ell}x\right) \le \left(\lambda^{\ell}L\right)^n \rho(x,\ldots,x)$$

for all $x, x_j \in A$ $(1 \le j \ne i \le n)$ and all positive integer λ.

Remark 4.2.7. If ρ is n-contractively subadditive and n-expansively superadditive separately, then ρ is contractively subhomogeneous of degree n and expansively superhomogeneous of degree n, respectively.

Assume that X and Y are linear spaces, $\mathcal{A}$ is a Lie C^*-algebra, $n \ge 2$ is a fixed positive integer and $n_o \in \mathbb{N}$ is a positive integer and suppose that $\mathbb{T}^1_{1/n_o} := \{e^{i\theta}; 0 \le \theta \le 2\pi/n_o\}$. For convenience, we use the following abbreviations for a given mapping $f : \mathcal{A} \to \mathcal{A}$,

$$\Delta_{\mu}f(x_1,\ldots,x_n) := \sum_{i=1}^{n} f\left(\mu x_i + \frac{1}{n-1}\sum_{j=1, j\ne i}^{n} \mu x_j\right) - 2\mu\sum_{i=1}^{n} f(x_i),$$
$$\Delta_{\alpha,\beta,\gamma}f(x,y) := \alpha f[x,y] - \beta[f(x),y] - \gamma[x,f(y)]$$

for all $x_1,\ldots,x_n \in \mathcal{A}$, all $\mu \in \mathbb{T}^1_{1/n_o}$ and $\alpha,\beta,\gamma \in \mathbb{C}$.

Lemma 4.2.8 (Najati and Ranjbari [50]). *A mapping $f : X \to Y$ satisfies the equation*

$$f\left(x_1 + \frac{x_2+x_3}{2}\right) + f\left(x_2 + \frac{x_1+x_3}{2}\right) + f\left(x_3 + \frac{x_1+x_2}{2}\right) = 2\big(f(x_1)+f(x_2)+f(x_3)\big) \tag{4.25}$$

for all $x_1,x_2,x_3 \in X$ if and only if f is additive.

It is noted that if $x_3 = 0$ in Eq. [4.25], we obtain theCauchy-Jensen equation

$$f\left(\frac{x_1+x_2}{2}\right) + f\left(x_1 + \frac{x_2}{2}\right) + f\left(\frac{x_1}{2} + x_2\right) = 2\big(f(x_1)+f(x_2)\big)$$

is equivalent to $f(x_1+x_2) = f(x_1)+f(x_2)$ for all $x_1,x_2 \in X$.

Lemma 4.2.9 (Eshaghi and Khodaei [48]). *A mapping $f : X \to Y$ satisfies Eq. [4.24] for all $x_1,\ldots,x_n \in X$ if and only if f is additive.*

Proof. The case $n = 2$ is trivial and therefore we assume that $n > 2$. Letting $x_1 = \cdots = x_n = 0$ in Eq. [4.24], we get $f(0) = 0$. Setting $x_1 = x$ and $x_2 = \cdots = x_n = 0$ in Eq. [4.24], we obtain $f\big((n-1)x\big) = (n-1)f(x)$ for all $x \in X$. So Eq. [4.24] may be rewritten to the form

$$\sum_{i=1}^{n} f\left((n-1)x_i + \sum_{j=1, j\neq i}^{n} x_j\right) = 2(n-1)\sum_{i=1}^{n} f(x_i) \tag{4.26}$$

for all $x_1, \ldots, x_n \in X$. Putting $x_1 = \cdots = x_n = x$ in Eq. [4.24], we obtain $f(2x) = 2f(x)$ for all $x \in X$. Setting $x_1 = \cdots = x_{n-2} = x$ and $x_{n-1} = x_n = 0$ in Eq. [4.26] and using $f(2x) = 2f(x)$, we get $f((n-2)x) = (n-2)f(x)$ for all $x \in X$. Putting $x_3 = \cdots = x_n = 0$ in Eq. [4.26], we get

$$(n-2)f(x_1+x_2) + f((n-1)x_1+x_2) + f(x_1+(n-1)x_2) = 2(n-1)(f(x_1)+f(x_2)) \tag{4.27}$$

for all $x_1, x_2 \in X$. Setting $x_1 = x$ and $x_2 = -x$ in Eq. [4.27] and using $f(0) = 0$, we obtain

$$f((n-2)x) + f((2-n)x) = 2(n-1)(f(x) + f(-x))$$

for all $x \in X$. So by $f((n-2)x) = (n-2)f(x)$ since $n > 2$, we infer that $f(-x) = -f(x)$ for all $x \in X$. Putting $x_4 = \cdots = x_n = 0$ in Eq. [4.26], we get

$$\begin{aligned}(n-3)f(x_1+x_2+x_3) &+ f((n-1)x_1+x_2+x_3) \\ &+ f(x_1+(n-1)x_2+x_3) + f(x_1+x_2+(n-1)x_3) \\ &= 2(n-1)(f(x_1)+f(x_2)+f(x_3))\end{aligned} \tag{4.28}$$

for all $x_1, x_2, x_3 \in X$. Setting $x_3 = -x_1 - x_2$ in Eq. [4.28] and using oddness of f, we obtain

$$(n-2)(f(x_1)+f(x_2)-f(x_1+x_2)) = (2n-2)(f(x_1)+f(x_2)-f(x_1+x_2))$$

for all $x \in X$. Therefore, f is an additive mapping. The converse implication is obvious. □

Lemma 4.2.10 (Eshaghi Gordji [51]). *Let $f : \mathcal{A} \to \mathcal{A}$ be an additive mapping such that $f(tx) = tf(x)$ for all $t \in \mathbb{T}^1_{1/n_0}$ and $x \in \mathcal{A}$. Then the mapping f is $\mathbb{C}$-linear.*

Theorem 4.2.11 (Eshaghi and Khodaei [48]). *Assume that there exist an expansively superhomogeneous mapping $\phi : \mathcal{A}^n = \overbrace{\mathcal{A} \times \cdots \times \mathcal{A}}^{n\text{-times}} \to (0, \infty)$ and a 2-expansively superhomogeneous $\psi : \mathcal{A}^2 \to (0, \infty)$ with a constant L such that a mapping $f : \mathcal{A} \to \mathcal{A}$ satisfies*

$$\left\| \Delta_\mu f(x_1, \ldots, x_n) \right\| \leq \phi(x_1, \ldots, x_n), \tag{4.29}$$

$$\left\| \Delta_{\alpha,\beta,\gamma} f(x, y) \right\| \leq \psi(x, y) \tag{4.30}$$

for all $x_1, \ldots, x_n, x, y \in \mathcal{A}$*, all* $\mu \in \mathbb{T}^1_{1/n_o}$ *and some* $\alpha, \beta, \gamma \in \mathbb{C}$*. Then there exists a unique Lie* (α, β, γ)*-derivation* $\mathfrak{D} : \mathcal{A} \to \mathcal{A}$*, which satisfies the Eq. [4.24] and the inequality*

$$\|f(x) - \mathfrak{D}(x)\| \leq \frac{L}{2n(1-L)} \quad \phi(\overbrace{x, \ldots, x}^{n\text{-times}}) \tag{4.31}$$

for all $x \in \mathcal{A}$.

Proof. Consider the set

$$\mathfrak{W} := \left\{ g : \mathcal{A} \to \mathcal{A}, \quad \sup_{x \in \mathcal{A}} \frac{\|g(x) - f(x)\|}{\phi\,(x, \ldots, x)} < \infty \right\}$$

and introduce the *metric* on $\mathfrak{W}$:

$$d(g, h) = \sup_{x \in \mathcal{A}} \frac{\|g(x) - h(x)\|}{\phi\,(x, \ldots, x)}.$$

Then $(\mathfrak{W}, d)$ is complete. Now we consider the mapping $\Lambda : \mathfrak{W} \to \mathfrak{W}$ defined by

$$(\Lambda g)(x) = 2g\left(\frac{x}{2}\right), \quad \text{for all } g \in \mathfrak{W} \text{ and } x \in \mathcal{A}.$$

Let $g, h \in \mathfrak{W}$ and let $C \in [0, \infty)$ be an arbitrary constant with $d(g, h) < C$. From the definition of d, we have

$$\frac{\|g(x) - h(x)\|}{\phi\,(x, \ldots, x)} \leq C$$

for all $x \in \mathcal{A}$. By the assumption and the last inequality, we have

$$\frac{\|(\Lambda g)(x) - (\Lambda h)(x)\|}{\phi\,(x, \ldots, x)} = \frac{2\left\|g\left(\frac{x}{2}\right) - h\left(\frac{x}{2}\right)\right\|}{\phi\,(x, \ldots, x)} \leq \frac{L\left\|g\left(\frac{x}{2}\right) - h\left(\frac{x}{2}\right)\right\|}{\phi\left(\frac{x}{2}, \ldots, \frac{x}{2}\right)} \leq LC$$

for all $x \in \mathcal{A}$. So

$$d(\Lambda g, \Lambda h) \leq Ld(g, h), \quad \forall g, h \in \mathfrak{W}.$$

This means Λ is a strictly contractive self-mapping of $\mathfrak{W}$, with the Lipschitz constant L.

Substituting $x_1, \ldots, x_n = x$ and $\mu = 1$ in the functional inequality [4.29], we obtain

$$\|nf(2x) - 2nf(x)\| \leq \phi\,(x, \ldots, x)$$

for all $x \in \mathcal{A}$. Thus

$$\frac{\left\|2f\left(\frac{x}{2}\right) - f(x)\right\|}{\phi(x, \ldots, x)} \leq \frac{L}{2n}$$

for all $x \in \mathcal{A}$. Hence, $d(\Lambda f, f) \leq \frac{L}{2n}$.

Due to Theorem 2.2.8, there exists a unique mapping $\mathfrak{D} \in \mathfrak{W}$ such that $\mathfrak{D}(2x) = 2\mathfrak{D}(x)$ for all $x \in \mathcal{A}$, ie, $\mathfrak{D}$ is a unique fixed point of Λ. Moreover,

$$\mathfrak{D}(x) = \lim_{m\to\infty} 2^m f\left(\frac{x}{2^m}\right) \tag{4.32}$$

for all $x \in \mathcal{A}$. Also

$$d(f, \mathfrak{D}) \leq \frac{1}{1-L} d(f, \Lambda f) \leq \frac{L}{2n(1-L)},$$

ie, inequality [4.31] holds true for all $x \in \mathcal{A}$.

In addition, it is clear from Eqs. [4.29] and [4.32] that the inequality

$$\begin{aligned}\left\|\Delta_\mu \mathfrak{D}(x_1, \ldots, x_n)\right\| &= \lim_{m\to\infty} 2^m \left\|\Delta_\mu f\left(\frac{x_1}{2^m}, \ldots, \frac{x_n}{2^m}\right)\right\| \\ &\leq \lim_{m\to\infty} 2^m \phi\left(\frac{x_1}{2^m}, \ldots, \frac{x_n}{2^m}\right) \\ &\leq \lim_{m\to\infty} L^m \phi(x_1, \ldots, x_n) = 0\end{aligned}$$

holds for all $x_1, \ldots, x_n \in \mathcal{A}$ and all $\mu \in \mathbb{T}^1_{1/n_o}$. So $\Delta_\mu \mathfrak{D}(x_1, \ldots, x_n) = 0$ for all $x_1, \ldots, x_n \in \mathcal{A}$ and all $\mu \in \mathbb{T}^1_{1/n_o}$. If we put $\mu = 1$ in the last equality, then $\mathfrak{D}$ is additive by Lemma 4.2.9. Letting $x_1 = x$ and $x_2 = \cdots = x_n = 0$ in last equality, we obtain $\mathfrak{D}(\mu x) = \mu\mathfrak{D}(x)$. Now, by using Lemma 4.2.10, we infer that the mapping $\mathfrak{D} \in \mathfrak{W}$ is $\mathbb{C}$-linear.

It follows from linearity of $\mathfrak{D}$ and Eq. [4.30] that

$$\begin{aligned}\left\|\Delta_{\alpha,\beta,\gamma}\mathfrak{D}(x, y)\right\| &= \lim_{m\to\infty} 4^m \left\|\Delta_{\alpha,\beta,\gamma} f\left(\frac{x}{2^m}, \frac{y}{2^m}\right)\right\| \leq \lim_{m\to\infty} 4^m \psi\left(\frac{x}{2^m}, \frac{y}{2^m}\right) \\ &\leq \lim_{m\to\infty} 4^m \left(\frac{L}{2}\right)^{2m} \psi(x, y) = 0\end{aligned}$$

for all $x, y \in \mathcal{A}$ and some $\alpha, \beta, \gamma \in \mathbb{C}$. So for some $\alpha, \beta, \gamma \in \mathbb{C}$,

$$\alpha\mathfrak{D}[x, y] = \beta[\mathfrak{D}(x), y] + \gamma[x, \mathfrak{D}(y)]$$

for all $x, y \in \mathcal{A}$. Thus the mapping $\mathfrak{D} \in \mathfrak{W}$ is a Lie (α, β, γ)-derivation.

Therefore, the mapping $\mathfrak{D} \in \mathfrak{W}$ is a unique Lie (α, β, γ)-derivation on Lie C^*-algebra $\mathcal{A}$ satisfying Eq. [4.31]. □

Theorem 4.2.12 (Eshaghi and Khodaei [48]). *Assume that there exists a contractively subhomogeneous mapping $\varphi : \mathcal{A}^{n+2} \to (0, \infty)$ with a constant L such that a mapping $f : \mathcal{A} \to \mathcal{A}$ satisfies*

$$\left\| \Delta_{\mu} f(x_1, \ldots, x_n) + \Delta_{\alpha,\beta,\gamma} f(x, y) \right\| \le \varphi(x_1, \ldots, x_n, x, y) \tag{4.33}$$

for all $x_1, \ldots, x_n, x, y \in \mathcal{A}$, all $\mu \in \mathbb{T}^1_{1/n_o}$ and some $\alpha, \beta, \gamma \in \mathbb{C}$. Then there exists a unique Lie (α, β, γ)-derivation $\mathfrak{D} : \mathcal{A} \to \mathcal{A}$ which satisfies the Eq. [4.24] and the inequality

$$\|f(x) - \mathfrak{D}(x)\| \le \frac{1}{2n(1-L)} \quad \varphi(x, \ldots, x, 0, 0) \tag{4.34}$$

for all $x \in \mathcal{A}$.

Proof. Substituting $x_1, \ldots, x_n = x$ and $\mu = 1$ in the functional inequality [4.33], we obtain

$$\|nf(2x) - 2nf(x)\| \le \varphi(x, \ldots, x, 0, 0) \tag{4.35}$$

for all $x \in \mathcal{A}$. We introduce the same definitions for $\mathfrak{W}$ and d as in the proof of Theorem 4.2.11 (by replacing ϕ by φ) such that $(\mathfrak{W}, d)$ becomes a complete metric space. Let $\Lambda : \mathfrak{W} \to \mathfrak{W}$ be the mapping defined by

$$(\Lambda g)(x) = \frac{1}{2} g(2x), \quad \text{for all } g \in \mathfrak{W} \text{ and } x \in \mathcal{A}.$$

One can show that $d(\Lambda g, \Lambda h) \le Ld(g, h)$ for any $g, h \in \mathfrak{W}$. It follows from Eq. [4.35] that

$$\frac{\left\| \frac{1}{2} f(2x) - f(x) \right\|}{\varphi(x, \ldots, x, 0, 0)} \le \frac{1}{2n}$$

for all $x \in \mathcal{A}$. Hence $d(\Lambda f, f) \le \frac{1}{2n}$. Due to Theorem 2.2.8, there exists a unique mapping $\mathfrak{D} : \mathcal{A} \to \mathcal{A}$ such that $\mathfrak{D}(2x) = 2\mathfrak{D}(x)$ for all $x \in \mathcal{A}$, ie, $\mathfrak{D}$ is a unique fixed point of Λ. Moreover,

$$\mathfrak{D}(x) = \lim_{m \to \infty} \frac{1}{2^m} f\left(2^m x\right)$$

for all $x \in \mathcal{A}$. Also,

$$d(f, \mathfrak{D}) \le \frac{1}{1-L} d(f, \Lambda f) \le \frac{1}{2n(1-L)}.$$

This implies that inequality [4.34] holds.

The remaining assertion goes through by a similar way to corresponding part of Theorem 4.2.11. □

4.3 Stability problem on C^*-ternary algebras

Ternary algebraic operations were considered in the 19th century by several mathematicians and physicists such as Cayley [52] who introduced the notions of cubic matrix, which in turn was generalized by Kapranov el al. [53]. The simplest example of such nontrivial ternary operation is given by the following composition rule:

$$\{a,b,c\}_{ijk} = \sum_{l,m,n} a_{nil} b_{ljm} c_{mkn} \quad (i,j,k,\ldots = 1,2,\ldots,\mathbb{N}).$$

Ternary structures and their generalization, the so-called n-ary structures, raise certain hopes in view of their applications in physics. Some significant physical applications are as follows (see [54, 55]):

(i) The algebra of "nonions" generated by two matrices,

$$\begin{pmatrix} 0 & 1 & 0 \\ 0 & 0 & 1 \\ 1 & 0 & 0 \end{pmatrix} \text{ and } \begin{pmatrix} 0 & 1 & 0 \\ 0 & 0 & \omega \\ \omega^2 & 0 & 0 \end{pmatrix}, \quad (\omega = e^{2\Pi i/3})$$

was introduced by Sylvester as a ternary analog of Hamilton's quaternions (cf. [56]).

(ii) A natural ternary composition of 4-vectors in the four-dimensional Minkowskian space time M_4 can be defined as an example of a ternary operation:

$$(X,Y,Z) \longmapsto U(X,Y,Z) \in M_4$$

with the resulting 4-vector U^μ defined via its components in a given coordinate system as follows:

$$U^\mu(X,Y,Z) = g^{\mu\sigma} \eta_{\sigma\nu\lambda\rho} X^\nu Y^\lambda Z^\rho, \quad \mu,\nu,\ldots = 0,1,2,3,$$

where $g^{\mu\sigma}$ is the metric tensor and $\eta_{\sigma\nu\lambda\rho}$ is the canonical volume element of M_4 (see [55])

(iii) The quark model inspired a particular brand of ternary algebraic systems. The "Nambu mechanics" are based on such structures (see [57, 58]). Quarks apparently couple by packs of 3.

There are also some applications, although still hypothetical, in the fractional quantum Hall effect, nonstandard statistics, supersymmetric theory, Yang-Baxter equation, etc. (cf. [53, 56, 59]). Following the terminology of Ref. [60], a nonempty set G with a ternary operation $[\cdot,\cdot,\cdot] : G^3 \to G$ is called a ternary groupoid and is denoted by $(G,[\cdot,\cdot,\cdot])$. The ternary groupoid $(G,[\cdot,\cdot,\cdot])$ is called commutative if

$[x_1, x_2, x_3] = [x_{\sigma(1)}, x_{\sigma(2)}, x_{\sigma(3)}]$ for all $x_1, x_2, x_3 \in G$ and all permutations σ of $\{1, 2, 3\}$. If a binary operation $\circ$ is defined on G such that $[x, y, z] = (x \circ y) \circ z$ for all $x, y, z \in G$, then we say that $[\cdot, \cdot, \cdot]$ is derived from $\circ$. We say that $(G, [\cdot, \cdot, \cdot])$ is a ternary semigroup if the operation $[\cdot, \cdot, \cdot]$ is associative, ie, if $[[x, y, z], u, v] = [x, [y, z, u], v] = [x, y, [z, u, v]]$ holds for all $x, y, z, u, v \in G$ (see [61]).

As it is extensively discussed in [62], the full description of a physical system $\mathbb{S}$ implies the knowledge of three basis ingredients: the set of the observables, the set of the states and the dynamics that describe the time evolution of the system by means of the time dependence of the expectation value of a given observable on a given statue. Originally the set of the observables was considered to be a C^*-algebra [63]. In many applications, however, this was shown not to be the must convenient choice, and the C^*-algebra was replaced by a von Neumann algebra, because the role of the representation turns out to be crucial mainly when long-range interactions are involved (see [64] and references therein). Here we used a different algebraic structure. A C^*-ternary algebra is a complex Banach space A, equipped with a ternary product $(x, y, z) \rightarrowtail [x, y, z]$ of A^3 into A, which is $\mathbb{C}$-linear in the outer variables, conjugate $\mathbb{C}$-linear in the middle variable, and associative in the sense that $[x, y, [z, w, v]] = [x, [w, z, y], v] = [[x, y, z], w, v]$, and satisfies $\|[x, y, z]\| \le \|x\| \cdot \|y\| \cdot \|z\|$ and $\|[x, x, x]\| = \|x\|^3$. If a C^*-ternary algebra $(A, [\cdot, \cdot, \cdot])$ has an identity, ie, an element $e \in A$ such that $x = [x, e, e] = [e, e, x]$ for all $x \in A$, then it is routine to verify that A, endowed with $x \circ y := [x, e, y]$ and $x^* := [e, x, e]$, is a unital C^*-algebra. Conversely, if $(A, \circ)$ is a unital C^*-algebra, then $[x, y, z] := x \circ y^* \circ z$ makes A into a C^*-ternary algebra. A $\mathbb{C}$-linear mapping $H : A \to B$ is called a C^*-ternary algebra homomorphism if

$$H([x, y, z]) = [H(x), H(y), H(z)]$$

for all $x, y, z \in A$.

4.3.1 Approximately J^*-homomorphisms

Eshaghi and Najati [41] proved the stability and superstability of J^*-homomorphisms between J^*-algebras for the generalized Jensen-type functional equation $f(\frac{x+y}{2}) + f(\frac{x-y}{2}) = f(x)$ using the alternative of fixed point. Throughout this section, assume that A, B are two J^*-algebras.

Theorem 4.3.1 (Eshaghi et al. [41]). *Let $f : A \to B$ be a mapping with $f(0) = 0$ for which there exists a function $\phi : A^3 \to [0, \infty)$ such that*

$$\left\| \mu f\left(\frac{x + zz^*z + y}{2}\right) + \mu f\left(\frac{x + zz^*z - y}{2}\right) - f(\mu x) - \mu f(z)f(z)^*f(z) \right\| \le \phi(x, y, z) \tag{4.36}$$

for all $\mu \in \mathbb{T} = \{\mu \in \mathbb{C} : |\mu| = 1\}$ and all $x, y, z \in A$. If there exists an $0 \le L < 1$ such that

$$\phi(x, y, z) \le 2L\phi\left(\frac{x}{2}, \frac{y}{2}, \frac{z}{2}\right) \tag{4.37}$$

for all $x, y, z \in A$, then there exists a unique J^*-homomorphism $h : A \to B$ such that

$$\|f(x) - h(x)\| \leq \frac{L}{1-L}\phi(x, 0, 0) \tag{4.38}$$

for all $x \in A$.

Proof. It follows from Eq. [4.37] that

$$2^{-j}\phi(2^j x, 2^j y, 2^j z) \leq L^j \phi(x, y, z)$$

for all $x, y, z \in A$. Hence,

$$\lim_{j\to\infty} 2^{-j}\phi(2^j x, 2^j y, 2^j z) = 0 \tag{4.39}$$

for all $x, y, z \in A$. Putting $\mu = 1, y = z = 0$ in Eq. [4.36], we obtain

$$\left\|2f\left(\frac{x}{2}\right) - f(x)\right\| \leq \phi(x, 0, 0) \tag{4.40}$$

for all $x \in A$. Hence,

$$\left\|\frac{1}{2}f(2x) - f(x)\right\| \leq \frac{1}{2}\phi(2x, 0, 0) \leq L\phi(x, 0, 0) \tag{4.41}$$

for all $x \in A$. Consider the set $X := \{g \mid g : A \to B, g(0) = 0\}$ and introduce the generalized metric on X:

$$d(h, g) := \inf\{C \in \mathbb{R}^+ : \|g(x) - h(x)\| \leq C\phi(x, 0, 0),\ \forall x \in A\}.$$

It is easy to show that (X, d) is complete. Now we define the linear mapping $J : X \to X$ by

$$J(h)(x) = \frac{1}{2}h(2x)$$

for all $x \in A$. It is easy to show that $d(J(g), J(h)) \leq Ld(g, h)$ for all $g, h \in X$ (see [65]). It follows from Eq. [4.41] that $d(f, J(f)) \leq L$. By using the alternative of fixed point, J has a unique fixed point in the set $X_1 := \{h \in X : d(f, h) < \infty\}$. So h satisfies $h(2x) = 2h(x)$ and

$$h(x) = \lim_{n\to\infty} \frac{1}{2^n}f(2^n x), \quad d(f, h) \leq \frac{1}{1-L}d(f, J(f))$$

for all $x \in A$. Therefore, $d(f, h) \leq \frac{L}{1-L}$. This implies inequality [4.38]. Put $z = 0$ in Eq. [4.36], it follows from the definition of J and Eq. [4.39] that

$$\left\|\mu h\left(\frac{x+y}{2}\right)+\mu h\left(\frac{x-y}{2}\right)-h(\mu x)\right\|$$
$$=\lim_{n\to\infty}\frac{1}{2^n}\|\mu f(2^{n-1}(x+y))+\mu f(2^{n-1}(x-y))-f(2^n\mu x)\|$$
$$\leq\lim_{n\to\infty}\frac{1}{2^n}\phi(2^nx,2^ny,0)=0$$

for all $x, y \in A$. So

$$\mu h\left(\frac{x+y}{2}\right)+\mu h\left(\frac{x-y}{2}\right)=h(\mu x)$$

for all $x, y \in A$. Put $u = \frac{x+y}{2}, v = \frac{x-y}{2}$ in the above equation, we get

$$\mu h(u)+\mu h(v)=h(\mu u+\mu v)$$

for all $u, v \in A$. Since $h(0) = 0$, h is additive and $h(\mu x) = \mu h(x)$ for all $\mu \in \mathbb{T}$ and all $x \in A$. Hence, h is $\mathbb{C}$-linear by Lemma 4.2.10. Setting $x = y = 0$ and $\mu = 1$ in Eq. [4.49], we have

$$\begin{aligned}\|h(zz^*z)-h(z)h(z)^*h(z)\| &= \left\|2h\left(\frac{zz^*z}{2}\right)-h(z)h(z)^*h(z)\right\|\\ &=\lim_{n\to\infty}\frac{1}{8^n}\left\|2f\left(\frac{8^nzz^*z}{2}\right)-h(2^nz)h(2^nz)^*h(2^nz)\right\|\\ &\leq\lim_{n\to\infty}\frac{1}{8^n}\phi(0,0,2^nz)\leq\lim_{n\to\infty}\frac{1}{2^n}\phi(0,0,2^nz)=0\end{aligned}$$

for all $z \in A$. Thus $h : A \to B$ is a J^*-homomorphism satisfying Eq. [4.38], as desired. □

We prove the following generalized Hyers-Ulam stability problem for J^*-homomorphisms on J^*-algebras.

Corollary 4.3.2 (Eshaghi et al. [41]). *Let $p \in (0, 1)$ and $\delta, \theta \in [0, \infty)$ be real numbers. Suppose $f : A \to B$ satisfies $f(0) = 0$ and*

$$\left\|\mu f\left(\frac{x+zz^*z+y}{2}\right)+\mu f\left(\frac{x+zz^*z-y}{2}\right)-f(\mu x)-f(z)f(z)^*f(z)\right\|$$
$$\leq\theta(\|x\|^p+\|y\|^p+\|z^p\|),$$

for all $\mu \in \mathbb{T}$ and all $x, y, z \in A$. Then there exists a unique J^*-homomorphism $h : A \to B$ such that

$$\|f(x)-h(x)\|\leq\frac{2^p\delta}{2-2^p}+\frac{2^p\theta}{2-2^p}\|x\|^p$$

for all $x \in A$.

Proof. Set $\phi(x,y,z) := \theta(\|x\|^p + \|y\|^p + \|z^p)$ for all $x, y, z \in A$. Then we get the desired result by $L = 2^{p-1}$ in Theorem 4.3.1. □

Remark 4.3.3 (Eshaghi et al. [41]). Let $f : A \to B$ be a mapping with $f(0) = 0$ for which there exists a function $\phi : A^3 \to [0, \infty)$ such that

$$\left\| \mu f\left(\frac{x + zz^*z + y}{2}\right) + \mu f\left(\frac{x + zz^*z - y}{2}\right) - f(\mu x) - \mu f(z)f(z)^*f(z) \right\| \\ \leq \phi(x,y,z)$$

for all $\mu \in \mathbb{T}$ and all $x, y, z \in A$. Let $0 < L < 1$ be a constant such that $2\phi(x,y,z) \leq L\phi(2x, 2y, 2z)$ for all $x, y, z \in A$. By a method similar to that of the proof of Theorem 4.3.1, one can show that there exists a unique J^*-homomorphism $h : A \to B$ satisfying

$$\|f(x) - h(x)\| \leq \frac{L}{1-L}\phi(x, 0, 0) \tag{4.42}$$

for all $x \in A$.

For the case $\phi(x,y,z) := \theta(\|x\|^p + \|y\|^p + \|z^p)$ (where θ is a nonnegative real number and $p > 1$), there exists a unique J^*-homomorphism $h : A \to B$ satisfying

$$\|f(x) - h(x)\| \leq \frac{2^p\theta}{2 - 2^p}\|x\|^p$$

for all $x \in A$.

The case in which $p = 1$ was excluded in Corollary 4.3.2 and Remark 4.3.3. Indeed, the results are not valid when $p = 1$. Here we use Gajda's example [10] to give a counter-example.

Proposition 4.3.4 (Eshaghi et al. [41]). *Let $\phi : \mathbb{C} \to \mathbb{C}$ be defined by*

$$\phi(x) := \begin{cases} x, & \text{for } |x| < 1, \\ 1, & \text{for } |x| \leq 1. \end{cases}$$

Consider the function $f : \mathbb{C} \to \mathbb{C}$ given by the formula

$$f(x) := \sum_{n=0}^{\infty} 2^{-n}\phi(2^n x).$$

Let

$$D_\mu f(x,y,z) = \mu f\left(\frac{x + z\bar{z}z + y}{2}\right) + \mu f\left(\frac{x + z\bar{z}z - y}{2}\right) - f(\mu x) - \mu f(z)\overline{f(z)}f(z)$$

for all $\mu \in \mathbb{T}$ and all $x, y, z \in A$. Then f satisfies

$$|D_\mu f(x,y,z)| \leq 36(|x| + |y| + |z|) \tag{4.43}$$

for all $\mu \in \mathbb{T}$ *and all* $x, y, z \in A$, *and the range of* $|f(x) - A(x)|/|x|$ *for* $x \neq 0$ *is unbounded for each additive function* $A : \mathbb{C} \to \mathbb{C}$.

Proof. It is clear that f is bounded by 2 on C. If $|x|+|y|+|z| = 0$ or $|x|+|y|+|z| \leq 1$, then

$$|D_\mu f(x, y, z)| \leq 14 \leq 14(|x| + |y| + |z|).$$

Now suppose that $0 < |x| + |y| + |z| < 1$. Then there exists an integer $k \geq 0$ such that

$$\frac{1}{2^{k+1}} \leq |x| + |y| + |z| \leq \frac{1}{2^k}. \tag{4.44}$$

Therefore,

$$2^m|x + z\bar{z}z + y|, 2^m|\mu x|, 2^m|z| < 1$$

for all $m = 0, 1, \ldots, k - 1$. From the definition of f and Eq. [4.44], we have

$$\begin{aligned}
|f(z)| &\leq k|z| + \sum_{n=k}^{\infty} 2^{-n}|\phi(2^n z)| \leq k|z| + \frac{2}{2^k}, \\
|D_\mu f(x, y, z)| &\leq |z|^3 + \frac{6}{2^k} + |f(z)|^3 \leq (k + k^3)|z|^3 + \frac{8}{2^k} + \frac{6k^2 + 12k}{4^k}|z| \\
&\leq \frac{k^3 + 6k^2 + 13k}{4^k}|z| + \frac{8}{2^k} \\
&\leq 20|z| + 16(|x| + |y| + |z|) \\
&\leq 36(|x| + |y| + |z|).
\end{aligned}$$

Therefore, f satisfies Eq. [4.43]. Let $A : \mathbb{C} \to \mathbb{C}$ be an additive function such that

$$|f(x) - A(x)| \leq \beta|x|$$

for all $x \in \mathbb{C}$. Then there exists a constant $c \in \mathbb{C}$ such that $A(x) = cx$ for all rational numbers x. So we have

$$|f(x)| \leq (\beta + |c|)|x| \tag{4.45}$$

for all rational numbers x. Let $m \in \mathbb{N}$ with $m > \beta + |c|$. If x is a rational number in $(0, 2^{1-m})$, then $2^n x \in (0, 1)$ for all $n = 0, 1, \ldots, m - 1$. So

$$f(x) \geq \sum_{n=0}^{m-1} 2^{-n}\phi(2^n x) = mx > (\beta + |c|)x,$$

which contradicts Eq. [4.45]. □

Now we establish the superstability of J^*-homomorphisms as follows.

Theorem 4.3.5 (Eshaghi et al. [41]). *Let $|r| > 1$, and let $f : A \to B$ be a mapping satisfying $f(rx) = rf(x)$ for all $x \in A$. Let $\phi : A^3 \to [0, \infty)$ be a mapping such that*

$$\begin{aligned} &\|\mu f(x + zz^*z + y) + \mu f(x - y) - 2f(\mu x) - \mu f(z)f(z)^*f(z)\| \\ &\quad \leq \phi(x, y, z) \end{aligned} \tag{4.46}$$

for all $\mu \in \mathbb{T}$ and all $x, y, z \in A$. If there exists a constant $0 < L < 1$ such that

$$\phi(x, y, z) \leq |r|L\phi\left(\frac{x}{r}, \frac{y}{r}, \frac{z}{r}\right)$$

for all $x, y, z \in A$, then f is a J^*-homomorphism.

Proof. By using equation $f(rx) = rf(x)$ and Eq. [4.46], we have $f(0) = 0$ and

$$\begin{aligned} &\|\mu f(x + y) + \mu f(x - y) - 2f(\mu x)\| \leq |r|^{-n}\phi(r^n x, r^n y, 0), \\ &\qquad \|f(zz^*z) - f(z)f(z)^*f(z)\| \leq |r|^{-3n}\phi(0, 0, r^n z) \end{aligned} \tag{4.47}$$

for all $x, y \in A$ and all integers n. It follows from $\phi(x, y, z) \leq |r|L\phi(\frac{x}{r}, \frac{y}{r}, \frac{z}{r})$ that

$$\lim_{n\to\infty} |r|^{-n}\phi(r^n x, r^n y, r^n z) = 0$$

for all $x, y, z \in A$. Hence, we get from Eq. [4.47] that

$$f(x + y) + \mu f(x - y) = 2f(\mu x), \quad f(zz^*z) = f(z)f(z)^*f(z)$$

for all $\mu \in \mathbb{T}$ and all $x, y, z \in A$. So f is additive and $f(\mu x) = \mu f(x)$ for all $\mu \in \mathbb{T}$ and all $x \in A$. By Lemma 4.2.10, f is $\mathbb{C}$-linear and we conclude that f is a J^*-homomorphism. □

Theorem 4.3.6 (Eshaghi et al. [41]). *Let $0 < |r| < 1$, and let $f : A \to B$ be a mapping satisfying $f(rx) = rf(x)$ for all $x \in A$. Let $\phi : A^3 \to [0, \infty)$ be a mapping satisfying Eq. [4.46]. If there exists a constant $0 < L < 1$ such that $|r|\phi(x, y, z) \leq L\phi(rx, ry, rz)$ for all $x, y, z \in A$, then f is a J^*-homomorphism.*

Corollary 4.3.7 (Eshaghi et al. [41]). *Let $0 < |r| \neq 1$, $p \in (0, 1)$, and $\delta, \theta \geq 0$ be real numbers. Suppose that $f : A \to B$ is a mapping satisfying $f(rx) = rf(x)$ for all $x \in A$ and inequality*

$$\begin{aligned} &\left\|\mu f\left(x + zz^*z + y\right) + \mu f(x - y) - 2f(\mu x) - \mu f(z)f(z)^*f(z)\right\| \\ &\quad \leq \delta + \theta\left(\|x\|^p + \|y\|^p + \|z\|^p\right) \end{aligned}$$

for all $\mu \in \mathbb{T}$ and all $x, y, z \in A$. Then f is a J^*-homomorphism.

Proof. Setting $\phi(x, y, z) := \delta + \theta(\|x\|^p + \|y\|^p + \|z\|^p)$ for all $x, y, z \in A$. For $|r| > 1$, $L = |r|^{p-1}$ and for $0 < |r| < 1$, let $L = |r|^{1-p}$. Then we get the desired result by Theorem 4.3.5 (for $|r| > 1$) and Theorem 4.3.6 (for $0 < |r| < 1$). □

Corollary 4.3.8 (Eshaghi et al. [41]). *Let $0 < |r| \neq 1$, $p > 1$, and $\theta \geq 0$ be real numbers. Suppose that $f : A \to B$ is a mapping satisfying $f(rx) = rf(x)$ for all $x \in A$ and inequality*

$$\begin{aligned}&\left\|\mu f\left(x+zz^*z+y\right)+\mu f(x-y)-2f(\mu x)-\mu f(z)f(z)^*f(z)\right\|\\&\quad\leq \theta\left(\|x\|^p+\|y\|^p+\|z\|^p\right)\end{aligned}$$

for all $\mu \in \mathbb{T}$ and all $x, y, z \in A$. Then f is a J^*-homomorphism.

4.3.2 The stability of J*-derivations

Let $\mathcal{H}$, $\mathcal{K}$ be two Hilbert spaces and let $\mathcal{B}(\mathcal{H},\mathcal{K})$ be the space of all bounded operators from $\mathcal{H}$ into $\mathcal{K}$. By a J^*-algebra, we mean a closed subspace $\mathcal{A}$ of $\mathcal{B}(\mathcal{H},\mathcal{K})$ such that $xx^*x \in \mathcal{A}$ whenever $x \in \mathcal{A}$. Many familiar spaces are J^*-algebras [66]. Of course J^*-algebras are not algebras in the ordinary sense. However, from one point of view they may be considered a generalization of C^*-algebras; see [66–68]. In particular, any Hilbert space may be thought of as a J^*-algebra identified with $\mathcal{L}(\mathcal{H},\mathbb{C})$. Also, any C^*-algebra in $\mathcal{B}(\mathcal{H})$ is a J^*-algebra. Other important examples of J^*-algebras are the so-called Cartan factors of types I, II, III, and IV. A J^*-derivation on a J^*-algebra $\mathcal{A}$ is defined to be a $\mathbb{C}$-linear mapping $d : \mathcal{A} \to \mathcal{A}$ such that

$$d(aa^*a) = d(a)a^*a + a(d(a))^*a + aa^*d(a)$$

for all $a \in \mathcal{A}$. In particular, every J^*-derivation on a C^*-algebra is a J^*-derivation.

Eshaghi et al. [69] investigated the stability and superstability of J^*-derivations in J^*-algebras for the generalized Jensen-type functional equation

$$rf\left(\frac{x+y}{r}\right)+rf\left(\frac{x-y}{r}\right)=2f(x). \tag{4.48}$$

Assume that $\mathcal{A}$ is a J^*-algebra. We have the following theorem in superstability of J^*-derivations.

Theorem 4.3.9 (Eshaghi et al. [69]). *Let $r, s \in (1,\infty)$, and let $D : \mathcal{A} \to \mathcal{A}$ be a mapping for which $D(sa) = sD(a)$ for all $a \in \mathcal{A}$. Suppose there exists a function $\phi : \mathcal{A}^3 \to [0,\infty)$ such that*

$$\lim_{n\to\infty} s^{-n}\phi(s^na, s^nb, s^nc) = 0,$$

$$\begin{aligned}&\left\|r\mu D\Big(\frac{a+b}{r}\Big)+r\mu D\Big(\frac{a-b}{r}\Big)-2D(\mu a)+D(cc^*c)-D(c)(c)^*c\right.\\&\quad\left.-cD(c)^*c-cc^*D(c)\right\|\\&\quad\leq \phi(a,b,c)\end{aligned} \tag{4.49}$$

for all $\mu \in \mathbb{T}$ and all $a, b, c \in \mathcal{A}$. Then D is a J^*-derivation.

Proof. Put $\mu = a = b = 0$ in Eq. [4.49]. Then

$$\begin{aligned}
&\|D(cc^*c) - D(c)c^*c - cD(c^*)c - cc^*D(c)\| \\
&= \frac{1}{s^{3n}}\|D((s^nc)(s^nc^*)(s^nc)) - D(s^nc)(s^nc^*)(s^nc) - (s^nc)D(s^nc^*)(s^nc) \\
&- (s^nc)(s^nc^*)D(s^nc)\| \le \frac{1}{s^{3n}}\phi(0,0,s^nc) \le \frac{1}{s^n}\phi(0,0,s^nc)
\end{aligned}$$

for all $c \in \mathcal{A}$. The right-hand side tends to zero as $n \to \infty$. So

$$D(cc^*c) = D(c)c^*c + cD(c^*)c + cc^*D(c)$$

for all $c \in \mathcal{A}$. Similarly, put $c = 0$ in Eq. [4.49], then

$$\begin{aligned}
&\left\| r\mu D\left(\frac{a+b}{r}\right) + r\mu D\left(\frac{a-b}{r}\right) - 2D(\mu a)\right\| \\
&\qquad = \frac{1}{s^n}\left\| r\mu D\left(\frac{s^na + s^nb}{r}\right) + r\mu D\left(\frac{s^na - s^nb}{r}\right) - 2D(\mu s^na)\right\| \\
&\qquad \le \frac{1}{s^n}\phi(s^na, s^nb, 0)
\end{aligned}$$

for all $a, b \in \mathcal{A}$. The right-hand side tends to zero as $n \to \infty$. So,

$$r\mu D\left(\frac{a+b}{r}\right) + r\mu D\left(\frac{a-b}{r}\right) = 2D(\mu a) \tag{4.50}$$

for all $\mu \in \mathbb{T}$ and all $a, b \in \mathcal{A}$. Put $\mu = 1$ in above equation. Then

$$rD\left(\frac{a+b}{r}\right) + rD\left(\frac{a-b}{r}\right) = 2D(a)$$

for all $a, b \in \mathcal{A}$. This means that D satisfies Eq. [4.48]. It is easy to show that D is additive. Putting $\mu = 1, b = 0$ in Eq. [4.50], we get

$$rD\left(\frac{a}{r}\right) = D(a)$$

for all $a \in \mathcal{A}$. Then, by Eq. [4.50], we obtain that

$$\mu D(a+b) + \mu D(a-b) = 2D(\mu a)$$

for all $\mu \in \mathbb{T}$ and all $a, b \in \mathcal{A}$. Replacing b with a in above equation, then by additivity of D, we obtain that $\mu D(a) = D(\mu a)$ for all $a \in \mathcal{A}$ and all $\mu \in \mathbb{T}$. So it is easy to show that D is $\mathbb{C}$-linear (see, for example, Theorem 1 of [47]). □

Theorem 4.3.10 (Eshaghi et al. [69]). *Let $r \in (1, \infty)$, and let $f : \mathcal{A} \to \mathcal{A}$ be a mapping with $f(0) = 0$ for which there exists a function $\phi : \mathcal{A}^3 \to [0, \infty)$ such that*

$$\Phi(a,b,c) := \sum_{0}^{\infty} 2^{-n}\phi(2^n a, 2^n b, 2^n c) < \infty,$$

$$\left\| r\mu f\left(\frac{a+b}{r}\right) + r\mu f\left(\frac{a-b}{r}\right) - 2f(\mu a) + f(cc^*c) - f(c)(c)^* \right.$$
$$\left. c - cf(c)^*c - cc^*f(c) \right\| \leq \phi(a,b,c) \tag{4.51}$$

for all $\mu \in \mathbb{T}$ and all $a, b, c \in \mathcal{A}$. Then there exists a unique J^*-derivation $D : \mathcal{A} \to \mathcal{A}$ such that

$$\|f(a) - D(a)\| \leq \Phi(a,a,0) \tag{4.52}$$

for all $a \in \mathcal{A}$.

Proof. Put $\mu = 1$ and $b = c = 0$ in Eq. [4.51]. It follows that

$$\|f(a) - r^{-1}f(ra)\| \leq \frac{1}{2}\phi(ra, 0, 0)$$

for all $a \in \mathcal{A}$. By induction, we can show that

$$\|f(a) - r^{-n}f(r^n a)\| \leq \frac{1}{2}\sum_{1}^{n}\phi(r^n a, 0, 0) \tag{4.53}$$

for all $a \in \mathcal{A}$. Replacing a by a^m in Eq. [4.53] and then dividing by r^m, we get

$$\|f(a^m) - r^{-n-m}f(r^{n+m}a)\| \leq \frac{1}{2r^m}\sum_{m}^{m+n}\phi(r^k a, 0, 0)$$

for all $a \in \mathcal{A}$. Hence, $\{r^{-n}f(r^n a)\}$ is a Cauchy sequence. Since A is complete,

$$D(a) := \lim_{n\to\infty} r^{-n}f(r^n a)$$

exists for all $a \in \mathcal{A}$. By using Eq. [4.49], one can show that

$$\left\| rD\left(\frac{a+b}{r}\right) + rD\left(\frac{a-b}{r}\right) - 2D(a) \right\|$$
$$= \lim_{n\to\infty}\frac{1}{r^n}\left\| rf\left(r^{n-1}(a+b)\right) + rf\left(r^{n-1}(a-b)\right) - 2f(r^n a) \right\|$$
$$\leq \lim_{n\to\infty}\frac{1}{r^n}\phi(r^n a, r^n b, 0) = 0$$

for all $a, b \in \mathcal{A}$. So

$$rD\left(\frac{a+b}{r}\right) + rD\left(\frac{a-b}{r}\right) = 2D(a)$$

for all $a, b \in \mathcal{A}$. Putting $U = \frac{a+b}{r}, V = \frac{a-b}{r}$ in the above equation, we get

$$r(D(U) + D(V)) = 2D\left(\frac{r(U+V)}{2}\right)$$

for all $U, V \in \mathcal{A}$. Hence, D is a Jensen-type function. On the other hand, we have

$$\|D(\mu a) - \mu D(a)\| = \lim_{n\to\infty} \frac{1}{r^n} \left\|f\left(\mu r^n a\right) - \mu f\left(r^n a\right)\right\| \leq \lim_{n\to\infty} \frac{1}{r^n}\phi\left(r^n a, r^n a, 0\right) = 0$$

for all $\mu \in \mathbb{T}$, and all $a \in \mathcal{A}$. So it is easy to show that D is $\mathbb{C}$-linear. It follows from Eq. [4.49] that

$$\begin{aligned}&\left\|D\left(cc^*c\right) - D(c)c^*c - cD(c^*)c - cc^*D(c)\right\| \\ &= \lim_{n\to\infty}\left\|\frac{1}{r^{3n}}f\left((r^nc)(r^nc^*)(r^nc)\right) - \frac{1}{r^n}f\left(r^nc\right)\frac{r^nc^*}{r^n}\frac{r^nc}{r^n} - \frac{r^nc}{r^n}\frac{1}{r^n}f\left(r^nc^*\right)\frac{r^nc}{r^n}\right. \\ &\left. -\frac{r^nc}{r^n}\frac{r^nc^*}{r^n}\frac{1}{r^n}f\left(r^nc\right)\right\| \leq \lim_{n\to\infty}\frac{1}{r^{3n}}\phi\left(0,0,r^nc\right) \leq \lim_{n\to\infty}\frac{1}{r^n}\phi\left(0,0,r^nc\right) = 0\end{aligned}$$

for all $c \in \mathcal{A}$. Thus $D : \mathcal{A} \to \mathcal{A}$ is a J^*-derivation satisfying Eq. [4.52], as desired. □

We consider the following Hyers-Ulam-Rassias stability problem for J^*-derivations on J^*-algebras.

Corollary 4.3.11 (Eshaghi et al. [69]). *Let $p \in (0,1), \theta \in [0,\infty)$ and $r \in (1,\infty)$ be real numbers. Suppose $f : A \to A$ satisfies*

$$\left\|r\mu f\left(\frac{a+b}{r}\right) + r\mu f\left(\frac{a-b}{r}\right) - 2f(\mu a) + f\left(cc^*c\right) - f(c)(c)^*c - cf(c)^*c - cc^*f(c)\right\| \\ \leq \theta\left(\|a\|^p + \|b\|^p + \|c^p\|\right),$$

for all $\mu \in \mathbb{T}$ and all $a, b, c \in \mathcal{A}$. Then there exists a unique J^*-derivation $D : \mathcal{A} \to \mathcal{A}$ such that

$$\|f(a) - D(a)\| \leq \frac{2^p\theta}{2^{p-1}-1}\|a\|^p$$

for all $a \in \mathcal{A}$.

Proof. It follows from Theorem 4.3.10 by putting $\phi(a,b,c) := \theta\,(\|a\|^p + \|b\|^p + \|c\|^p)$ for all $a,b,c \in \mathcal{A}$. □

In the next theorem, we investigate the stability of J^*-derivations by using the fixed point alternative.

Theorem 4.3.12 (Eshaghi et al. [69]). *Let $r \in (1,\infty)$ be a real number. Let $f : \mathcal{A} \to \mathcal{A}$ be a mapping for which there exists a function $\phi : \mathcal{A}^3 \to [0,\infty)$ such that*

$$\left\| r\mu f\left(\frac{a+b}{r}\right) + r\mu f\left(\frac{a-b}{r}\right) - 2f(\mu a) + f\left(cc^*c\right) - f(c)(c)^* \right.$$
$$\left. c - cf(c)^*c - cc^*f(c) \right\| \le \phi(a,b,c) \tag{4.54}$$

for all $\mu \in \mathbb{T}$ and all $a,b,c \in \mathcal{A}$. If there exists an $L < 1$ such that

$$\phi(a,b,c) \le rL\phi\left(\frac{a}{r},\frac{b}{r},\frac{c}{r}\right) \tag{4.55}$$

for all $a,b,c \in \mathcal{A}$, then there exists a unique J^*-derivation $D : \mathcal{A} \to \mathcal{A}$ such that

$$\|f(a) - D(a)\| \le \frac{L}{1-L}\phi(a,0,0) \tag{4.56}$$

for all $a \in \mathcal{A}$.

Proof. Putting $\mu = 1, b = c = 0$ in Eq. [4.54], we obtain

$$\left\| 2rf\left(\frac{a}{r}\right) - 2f(a) \right\| \le \phi(a,0,0)$$

for all $a \in \mathcal{A}$. Hence,

$$\left\| \frac{1}{r}f(ra) - f(a) \right\| \le \frac{1}{2r}\phi(ra,0,0) \le L\phi(ra,0,0) \tag{4.57}$$

for all $a \in \mathcal{A}$.

Consider the set $X := \{g \mid g : A \to \mathcal{A}\}$ and introduce the generalized metric on X:

$$d(h,g) := inf\{C \in \mathbb{R}^+ : \|g(a) - h(a)\| \le C\phi(a,0,0) \forall a \in \mathcal{A}\}.$$

It is easy to show that (X,d) is complete. Now we define the linear mapping $J : X \to X$ by

$$J(h)(a) = \frac{1}{r}h(ra)$$

for all $a \in \mathcal{A}$. We have

$$d(J(g), J(h)) \leq Ld(g, h)$$

for all $g, h \in X$.

It follows from Eq. [4.57] that

$$d(f, J(f)) \leq L.$$

By the alternative of fixed point (Theorem 2.2.8), J has a unique fixed point in the set $X_1 := \{h \in X : d(f, h) < \infty\}$. Let D be the fixed point of J. D is the unique mapping satisfying

$$D(ra) = rD(a)$$

for all $a \in \mathcal{A}$ such that there exists $C \in (0, \infty)$ satisfying

$$\|D(a) - f(a)\| \leq C\phi(a, 0, 0)$$

for all $a \in \mathcal{A}$. On the other hand we have $\lim_{n\to\infty} d(J^n(f), D) = 0$. It follows that

$$\lim_{n\to\infty} \frac{1}{2^n} f(2^n a) = D(a)$$

for all $a \in \mathcal{A}$. It follows from $d(f, h) \leq \frac{1}{1-L} d(f, J(f))$, that

$$d(f, h) \leq \frac{L}{1-L}.$$

This implies the inequality [4.56].

It follows from Eq. [4.55] that

$$\lim_{j\to\infty} r^{-j}\phi\left(r^j a, r^j b, r^j c\right) = 0$$

for all $a, b, c \in \mathcal{A}$.

By the same reasoning as in the proof of Theorem 4.3.10, one can show that the mapping $D : \mathcal{A} \to \mathcal{A}$ is a J^*-derivation satisfying Eq. [4.56], as desired. □

We consider the proof of the following Hyers-Ulam-Rassias stability problem for J^*-derivations on J^*-algebras.

Corollary 4.3.13 (Eshaghi et al. [69]). *Let $p \in (0,1), \theta \in [0,\infty)$ be real numbers. Suppose $f : A \to A$ satisfies*

$$\left\| r\mu f\left(\frac{a+b}{r}\right) + r\mu f\left(\frac{a-b}{r}\right) - 2f(\mu a) + f\left(cc^*c\right) - f(c)(c)^*c - cf(c)^*c - cc^*f(c) \right\| \\ \leq \theta\left(\|a\|^p + \|b\|^p + \|c^p\|\right)$$

for all $\mu \in \mathbb{T}$ and all $a, b, c \in \mathcal{A}$. Then there exists a unique J^*-derivation $D : \mathcal{A} \to \mathcal{A}$ such that

$$\|f(a) - D(a)\| \leq \frac{2^p\theta}{2-2^p}\|a\|^p$$

for all $a \in \mathcal{A}$.

Proof. Set $\phi(a,b,c) := \theta\left(\|a\|^p + \|b\|^p + \|c^p\right)$ all $a, b, c \in \mathcal{A}$. Letting $L = 2^{p-1}$, we get the desired result. □

Now we establish the superstability of J^*-derivations by using the alternative of fixed point.

Theorem 4.3.14 (Eshaghi et al. [69]). *Let $s > 1$, and let $f : \mathcal{A} \to \mathcal{A}$ be a mapping satisfying $f(sx) = sf(x)$ for all $x \in \mathcal{A}$. Let $\phi : \mathcal{A}^3 \to [0,\infty)$ be a mapping satisfying Eq. [4.54]. If there exists an $L < 1$ such that*

$$\phi(x,y,z) \leq rL\phi\left(\frac{x}{r}, \frac{y}{r}, \frac{z}{r}\right)$$

for all $x, y, z \in \mathcal{A}$, then f is a J^*-derivation.

Proof. It is similar to the proof of Theorem 4.3.9. □

Corollary 4.3.15 (Eshaghi et al. [69]). *Let $r, p \in (0,1), \theta \in [0,\infty)$ be real numbers. Suppose $f : A \to A$ is a function satisfying $f(rx) = rf(x)$ for all $x \in \mathcal{A}$. Let $\phi : \mathcal{A}^3 \to [0,\infty)$ be a mapping satisfying Eq. [4.54]. Then f is a J^*-derivation.*

Proof. Set $\phi(x,y,z) := \theta\left(\|x\|^p + \|y\|^p + \|z\|^p\right)$ all $x, y, z \in \mathcal{A}$. Letting $L = 2^{p-1}$, we get the desired result. □

4.4 General solutions of some functional equations

Let R be a unital ring and M be a (two-sided) module on R. A function $M : R \to R$ is said to be a multiplicative (3-multiplicative) function if

$$M(ab) = M(a)M(b) \quad or \quad (M(abc) = M(a)M(b)M(c))$$

for all $a, b, c \in R$. It is easy to see that every multiplicative function is 3-multiplicative, and the converse is not true. A mapping $M : R \to R$ is said to be Jordan multiplicative if

$$M(ab+ba)=M(a)M(b)+M(b)M(a)$$

for all $a,b\in R$. A function $D:R\to M$ is called multiplicative derivation if

$$D(ab)=D(a)b+aD(b)$$

for all $a,b\in R$ and $D:R\to M$ is said to be a Jordan multiplicative derivation if

$$D(ab+ba)=D(a)b+aD(b)+D(b)a+bD(a)$$

for all $a,b\in R$. The functional equation

$$f(x+y+xy)=f(x)+f(y)+f(x)f(y)$$

is called Pompeiu's functional equation [70, 71]. In this section, we show that if $f:R\to R$ is a Pompeiu function, then f satisfies the functional equation

$$f(x+y+z+xy+xz+yz+xyz)=f(x)+f(y)+f(z)+f(x)f(y)+f(x)f(z)+f(y)f(z)+f(x)f(y)f(z). \tag{4.58}$$

Moreover, we show that a function $f:R\to R$ satisfies

$$f(x+y+xy+yx)=f(x)+f(y)+f(x)f(y)+f(y)f(x) \tag{4.59}$$

if and only if there exists a Jordan multiplicative function $M:R\to R$ such that

$$f(x)=M\left(x+\frac{1}{2}\right)-\frac{1}{2}$$

for all $x\in R$. This functional equation can be considered as a Jordan type of Pompeiu's functional equation.

We show that $f:R\to M$ satisfies

$$f(x+y+xy)=f(x)+f(y)+f(x)y+xf(y),$$

if and only if there exists a multiplicative derivation $D:R\to M$ such that

$$f(x)=D(x+1)$$

for all $x\in R$. Finally we show that $f:R\to M$ satisfies

$$f(x+y+xy+yx)=f(x)+f(y)+f(x)y+xf(y)+f(y)x+yf(x),$$

if and only if there exists a Jordan multiplicative derivation $D : R \to M$ such that

$$f(x) = D\left(x + \frac{1}{2}\right)$$

for all $x \in R$.

Let $\mathcal{X}$ be a Banach bimodule over a Banach algebra $\mathcal{A}$. Recall that $\mathcal{X} \oplus_1 \mathcal{A}$ is a Banach algebra equipped with the ℓ_1-norm

$$\|(x, a)\| = \|x\| + \|a\| \quad (a \in \mathcal{A}, x \in \mathcal{X})$$

and the product

$$(x_1, a_1)(x_2, a_2) = (x_1 \cdot a_2 + a_1 \cdot x_2, a_1 a_2) \quad (a_1, a_2 \in \mathcal{A}, x_1, x_2 \in \mathcal{X}).$$

The algebra $\mathcal{X} \oplus_1 \mathcal{A}$ is called a module extension Banach algebra. We refer the reader to [43] for more information on Banach modules and to [44, 72] for details on module extensions. We use this notion to find an equivalent functional equation for Pompeiu's functional equation.

4.4.1 A sufficient condition for Pompeiu's functional equation

It is known that $f : R \to R$ is a Pompeiu's function if and only if there exists a multiplicative function $M : R \to R$ such that

$$f(x) = M(x + 1) - 1,$$

where 1 is the unite element of R. Then by part (ii) of the following theorem, one can see that Eq. [4.58] is a sufficient condition for Pompeiu functional equation. We shall show that Eq. [4.58] is not a necessary condition for Pompeiu's functional equation.

Theorem 4.4.1.

(i) *If $f : R \to R$ holds in functional equation* [4.58], *then the function $M : R \to R$ defined by $M(x) = f(x - 1) + 1 (x \in R)$ is a 3-multiplicative function on R.*
(ii) *If $M : R \to R$ is a multiplicative function, then the function $f : R \to R$ defined by $f(x) = M(x + 1) - 1 (x \in R)$ holds in Eq. [4.58].*

Proof.

(i) Let $f : R \to R$ holds in Eq. [4.58] and put $M(x) = f(x - 1) + 1$ for all $x \in R$. We have

$$\begin{aligned} M(x)M(y)M(z) &= (f(x-1)+1)(f(y-1)+1)(f(z-1)+1) \\ &= f(x-1)f(y-1)f(z-1) + f(x-1)f(y-1) \\ &\quad + f(x-1)f(z-1) + f(y-1)f(z-1) \end{aligned}$$

$$\begin{aligned}&+f(x-1)+f(y-1)+f(z-1)+1\\&=f[(x-1)(y-1)(z-1)+(x-1)(y-1)\\&\quad+(x-1)(z-1)+(y-1)(z-1)\\&\quad+(x-1)+(y-1)+(z-1)]+1\\&=f(xyz-1)+1\\&=M(xyz)\end{aligned}$$

for all $x, y, z \in R$. This means that $M : R \to R$ is 3-multiplicative.

(ii) Let $M : R \to R$ be a multiplicative function and define $f : R \to R$ by $f(x) = M(x+1) - 1$ for all $x \in R$. We know that f is a Pompeiu function. Then we have

$$\begin{aligned}f(xy+x+y)&=f(x)f(y)+f(x)+f(y),\\f(xz+x+z)&=f(x)f(z)+f(x)+f(z),\\f(yz+y+z)&=f(y)f(z)+f(y)+f(z)\end{aligned}$$

for all $x, y, z \in R$. It follows that

$$\begin{aligned}f(x)f(y)f(z)&=(M(x+1)-1)(M(y+1)-1)(M(z+1)-1)\\&=M(x+1)M(y+1)M(z+1)-M(x+1)M(y+1)\\&\quad-M(x+1)M(z+1)-M(y+1)M(z+1)+M(x+1)\\&\quad+M(y+1)+M(z+1)-1\\&=M[(x+1)(y+1)(z+1)]-M[(x+1)(y+1)]\\&\quad-M[(x+1)(z+1)]-M[(y+1)(z+1)]+M(x+1)\\&\quad+M(y+1)+M(z+1)-1\\&=[M(xyz+xy+xz+yz+x+y+z+1)-1]\\&\quad-[M(xy+x+y+1)-1]-[M(xz+x+z+1)-1]\\&\quad-[M(yz+y+z+1)-1]+M(x+1)-1\\&\quad+M(y+1)-1+M(z+1)-1\\&=[f(xyz+xy+xz+yz+x+y+z)]-[f(xy+x+y)]\\&\quad-[f(xz+x+z)]-[f(yz+y+z)]+f(x)+f(y)+f(z)\\&=[f(xyz+xy+xz+yz+x+y+z)]-[f(x)f(y)+f(x)+f(y)]\\&\quad-[f(x)f(z)+f(x)+f(z)]-[f(y)f(z)+f(y)+f(z)]\\&\quad+f(x)+f(y)+f(z)\end{aligned}$$

for all $x, y, z \in R$. It follows that f holds in Eq. [4.58]. □

Note that the converse of (ii) is not true. To this end, define $f : \mathbb{R} \to \mathbb{R}$ by $f(x) = -x - 2$. It is easy to see that f satisfies Eq. [4.58] and that f is not a Pompeiu function.

4.4.2 Functional equation $f(x+y+xy+yx)=f(x)+f(y)+f(x)f(y)+f(y)f(x)$

Let R be a unital ring such that $2=1+1$ is invertible. Then we find a solution for functional equation [4.59] by using the Jordan multiplicative mappings on R.

Theorem 4.4.2.

(i) *If $f:R\to R$ holds in functional equation* [4.59], *then the function $M:R\to R$ defined by $M(x):=f\left(x-\frac{1}{2}\right)+\frac{1}{2}$, $x\in R$, is a Jordan multiplicative function on R.*

(ii) *If $M:R\to R$ is a Jordan multiplicative function and define $f:R\to R$ by $f(x):=M\left(x+\frac{1}{2}\right)-\frac{1}{2}$, $x\in R$, then f holds in Eq. [4.59].*

Proof.

(i) Let $f:R\to R$ holds in Eq. [4.59] and put $M(x):=f\left(x-\frac{1}{2}\right)+\frac{1}{2}$ for all $x\in R$. We have

$$\begin{aligned}
&M(x)M(y)+M(y)M(x)\\
&=\left(f\left(x-\frac{1}{2}\right)+\frac{1}{2}\right)\left(f\left(y-\frac{1}{2}\right)+\frac{1}{2}\right)\\
&\quad+\left(f\left(y-\frac{1}{2}\right)+\frac{1}{2}\right)\left(f\left(x-\frac{1}{2}\right)+\frac{1}{2}\right)\\
&=\left(f\left(x-\frac{1}{2}\right)\right)\left(f\left(y-\frac{1}{2}\right)\right)+\frac{1}{2}f\left(x-\frac{1}{2}\right)+\frac{1}{2}f\left(y-\frac{1}{2}\right)+\frac{1}{4}\\
&\quad+\left(f\left(y-\frac{1}{2}\right)\right)\left(f\left(x-\frac{1}{2}\right)\right)+\frac{1}{2}f\left(y-\frac{1}{2}\right)+\frac{1}{2}f\left(x-\frac{1}{2}\right)+\frac{1}{4}\\
&=\left(f\left(x-\frac{1}{2}\right)\right)\left(f\left(y-\frac{1}{2}\right)\right)+\left(f\left(y-\frac{1}{2}\right)\right)\left(f\left(x-\frac{1}{2}\right)\right)\\
&\quad+f\left(x-\frac{1}{2}\right)+f\left(y-\frac{1}{2}\right)+\frac{1}{2}\\
&=f\left[\left(x-\frac{1}{2}\right)\left(y-\frac{1}{2}\right)+\left(y-\frac{1}{2}\right)\left(x-\frac{1}{2}\right)+\left(x-\frac{1}{2}\right)\right.\\
&\quad\left.+\left(y-\frac{1}{2}\right)\right]+\frac{1}{2}\\
&=f\left[xy+yx-\frac{1}{2}\right]+\frac{1}{2}\\
&=M(xy+yx)
\end{aligned}$$

for all $x,y\in R$. This means that $M:R\to R$ is a Jordan multiplicative function.

(ii) Let $M:R\to R$ be a Jordan multiplicative function and define $f:R\to R$ by $f(x):=M\left(x+\frac{1}{2}\right)-\frac{1}{2}$ for all $x\in R$. We show that f satisfies Eq. [4.59]. To this end, we have

$$\begin{aligned}
&f(x)f(y)+f(y)f(x)\\
&=\left(M\left(x+\frac{1}{2}\right)-\frac{1}{2}\right)\left(M\left(y+\frac{1}{2}\right)-\frac{1}{2}\right)
\end{aligned}$$

$$
\begin{aligned}
&+\left(M\left(y+\frac{1}{2}\right)-\frac{1}{2}\right)\left(M\left(x+\frac{1}{2}\right)-\frac{1}{2}\right)\\
&=\left(M\left(x+\frac{1}{2}\right)\right)\left(M\left(y+\frac{1}{2}\right)\right)-\frac{1}{2}M\left(x+\frac{1}{2}\right)-\frac{1}{2}M\left(y+\frac{1}{2}\right)+\frac{1}{4}\\
&+\left(M\left(y+\frac{1}{2}\right)\right)\left(M\left(x+\frac{1}{2}\right)\right)-\frac{1}{2}M\left(y+\frac{1}{2}\right)-\frac{1}{2}M\left(y+\frac{1}{2}\right)+\frac{1}{4}\\
&=\left(M\left(x+\frac{1}{2}\right)\right)\left(M\left(y+\frac{1}{2}\right)\right)+\left(M\left(y+\frac{1}{2}\right)\right)\left(M\left(x+\frac{1}{2}\right)\right)\\
&-M\left(x+\frac{1}{2}\right)-M\left(y+\frac{1}{2}\right)+\frac{1}{2}\\
&=M\left[\left(x+\frac{1}{2}\right)\left(y+\frac{1}{2}\right)+\left(y+\frac{1}{2}\right)\left(x+\frac{1}{2}\right)\right]\\
&-M\left(x+\frac{1}{2}\right)-M\left(y+\frac{1}{2}\right)+\frac{1}{2}\\
&=M\left[\left(x+\frac{1}{2}\right)\left(y+\frac{1}{2}\right)+\left(y+\frac{1}{2}\right)\left(x+\frac{1}{2}\right)\right]-\frac{1}{2}\\
&-\left[M\left(x+\frac{1}{2}\right)-\frac{1}{2}\right]-\left[M\left(y+\frac{1}{2}\right)-\frac{1}{2}\right]\\
&=M\left[xy+yx+x+y+\frac{1}{2}\right]-\frac{1}{2}\\
&-\left[M\left(x+\frac{1}{2}\right)-\frac{1}{2}\right]-\left[M\left(y+\frac{1}{2}\right)-\frac{1}{2}\right]\\
&=f(xy+x+y)-f(x)-f(y)
\end{aligned}
$$

□

for all $x, y \in R$. This means that f satisfies Eq. [4.2].

4.4.3 Functional equation $f(x+y+xy)=f(x)+f(y)+f(x)y+xf(y)$

Let R be a unital ring, M be a module on R and $f : R \to M$ be a function. Then we find a solution for functional equation

$$f(x+y+xy)=f(x)+f(y)+f(x)y+xf(y) \tag{4.60}$$

by using the multiplicative derivations from R into M. Moreover, by using Pompeiu's functional equation, we find an equivalent for Eq. [4.60].

Theorem 4.4.3.

(i) *If $f : R \to M$ holds in functional equation* [4.60], *then the function $D : R \to M$ define by $D(x) := f(x-1)$ for all $x \in R$ is a multiplicative derivation on R.*

(ii) *If $D : R \to M$ is a multiplicative derivation and define $f : R \to M$ by $f(x) := D(x+1)$ for all $x \in R$, then f holds in Eq. [4.60].*

Proof.

(i) Let $f : R \to M$ holds in Eq. [4.60] and put $D(x) := f(x-1)$ for all $x \in R$. We have

$$\begin{aligned} D(x)y + xD(y) &= (f(x-1))y + x(f(y-1)) \\ &= (f(x-1))y + x(f(y-1)) + f(x-1) - f(x-1) \\ &\quad + f(y-1) - f(y-1) \\ &= (f(x-1))(y-1) + (x-1)(f(y-1)) + f(x-1) + f(y-1) \\ &= f[(x-1) + (y-1) + (x-1)(y-1)] \\ &= f(xy-1) \\ &= D(xy) \end{aligned}$$

for all $x, y \in R$. This means that $D : R \to M$ is a multiplicative derivation.

(ii) Let $D : R \to M$ be a multiplicative derivation and define $f : R \to M$ by $f(x) := D(x+1)$ for all $x \in R$. We show that f satisfies Eq. [4.60]. To this end, we have

$$\begin{aligned} f(x)y + xf(y) &= D(x+1)y + xD(y+1) \\ &= D(x+1)(y+1) + (x+1)D(y+1) - D(x+1) - D(y+1) \\ &= D[(x+1)(y+1)] - D(x+1) - D(y+1) \\ &= D[xy + x + y + 1] - D(x+1) - D(y+1) \\ &= f(xy + x + y) - f(x) - f(y) \end{aligned}$$

for all $x, y \in R$. This means that f satisfies Eq. [4.3].

□

Theorem 4.4.4. *The mapping $f : \mathcal{A} \to \mathcal{X}$ holds in functional equation* [4.60], *if and only if the mapping $\varphi_f : \mathcal{A} \to \mathcal{X} \oplus_1 \mathcal{A}$ defined by $\varphi_f(a) := (f(a), a)$, is a Pompeiu function. Moreover, f is continuous if and only if φ_f is continuous.*

Proof. We have

$$\varphi_f(a+b+ab) = (f(a+b+ab), a+b+ab) = (f(a)+f(b)+f(a)b+af(b), a+b+ab)$$

for all $a, b \in \mathcal{A}$. On the other hand, we have

$$\varphi_f(a)\varphi_f(b) = (f(a), a)(f(b), b) = (f(a)b + af(b), ab)$$

for all $a, b \in \mathcal{A}$. It follows that

$$\begin{aligned} \varphi_f(a+b+ab) &= (f(a), a) + (f(b), b) + (f(a)b + af(b), ab) \\ &= \varphi_f(a) + \varphi_f(b) + \varphi_f(a)\varphi_f(b) \end{aligned}$$

for all $a, b \in \mathcal{A}$. This means that φ_f is a Pompeiu function. The continuity property follows easily from definitions. □

Let R be a unital ring such that $2 = 1 + 1$ is invertible and let M be a module on R. Then we find a solution for functional equation

$$f(x+y+xy+yx) = f(x) + f(y) + f(x)y + xf(y) + f(y)x + yf(x) \tag{4.61}$$

by using Jordan multiplicative derivations from R into M.

Theorem 4.4.5.

(i) *If $f : R \to M$ holds in functional equation* [4.61], *then the function $D : R \to M$ define by $D(x) := f\left(x - \frac{1}{2}\right)$ for all $x \in R$ is a Jordan multiplicative derivation on R.*

(ii) *If $D : R \to M$ is a Jordan multiplicative derivation and define $f : R \to M$ by $f(x) := D\left(x + \frac{1}{2}\right)$ for all $x \in R$, then f holds in Eq. [4.61].*

Proof.

(i) Let $f : R \to M$ holds in Eq. [4.61] and put $D(x) := f\left(x - \frac{1}{2}\right)$ for all $x \in R$. We have

$$\begin{aligned}
&D(x)y + xD(y) + D(y)x + yD(x) \\
&= \left(f\left(x - \frac{1}{2}\right)\right)y + x\left(f\left(y - \frac{1}{2}\right)\right) \\
&\quad + \left(f\left(y - \frac{1}{2}\right)\right)x + y\left(f\left(x - \frac{1}{2}\right)\right) \\
&= \left(f\left(x - \frac{1}{2}\right)\right)\left(y - \frac{1}{2}\right) + \left(x - \frac{1}{2}\right)\left(f\left(y - \frac{1}{2}\right)\right) + \frac{1}{2}f\left(x - \frac{1}{2}\right) \\
&\quad + \frac{1}{2}f\left(y - \frac{1}{2}\right) + \left(f\left(y - \frac{1}{2}\right)\right)\left(x - \frac{1}{2}\right) + \left(y - \frac{1}{2}\right)\left(f\left(x - \frac{1}{2}\right)\right) \\
&\quad + \frac{1}{2}f\left(y - \frac{1}{2}\right) + \frac{1}{2}f\left(x - \frac{1}{2}\right) \\
&= f\left(x - \frac{1}{2}\right) + f\left(y - \frac{1}{2}\right) + \left(f\left(x - \frac{1}{2}\right)\right)\left(y - \frac{1}{2}\right) \\
&\quad + \left(x - \frac{1}{2}\right)\left(f\left(y - \frac{1}{2}\right)\right) + \left(f\left(y - \frac{1}{2}\right)\right)\left(x - \frac{1}{2}\right) \\
&\quad + \left(y - \frac{1}{2}\right)\left(f\left(x - \frac{1}{2}\right)\right) \\
&= f\left[\left(x - \frac{1}{2}\right) + \left(y - \frac{1}{2}\right) + \left(x - \frac{1}{2}\right)\left(y - \frac{1}{2}\right) + \left(y - \frac{1}{2}\right)\left(x - \frac{1}{2}\right)\right] \\
&= f(xy + yx - \frac{1}{2}) \\
&= D(xy + yx)
\end{aligned}$$

for all $x, y \in R$. This means that $D : R \to M$ is a Jordan multiplicative derivation.

(ii) Let $D : R \to M$ be a Jordan multiplicative derivation and define $f : R \to M$ by $f(x) := D\left(x + \frac{1}{2}\right)$ for all $x \in R$. We show that f satisfies Eq. [4.61]. To this end, we have

$$\begin{aligned}
&f(x)y + xf(y) + f(y)x + yf(x) \\
&= D\left(x+\frac{1}{2}\right)y + xD\left(y+\frac{1}{2}\right) \\
&\quad + D\left(y+\frac{1}{2}\right)x + yD\left(x+\frac{1}{2}\right) \\
&= D\left(x+\frac{1}{2}\right)\left(y+\frac{1}{2}\right) + \left(x+\frac{1}{2}\right)D\left(y+\frac{1}{2}\right) + D\left(y+\frac{1}{2}\right)\left(x+\frac{1}{2}\right) \\
&\quad + \left(y+\frac{1}{2}\right)D\left(x+\frac{1}{2}\right) - D\left(x+\frac{1}{2}\right) - D\left(y+\frac{1}{2}\right) \\
&= D\left[\left(x+\frac{1}{2}\right)\left(y+\frac{1}{2}\right) + \left(y+\frac{1}{2}\right)\left(x+\frac{1}{2}\right)\right] - D\left(x+\frac{1}{2}\right) - D\left(y+\frac{1}{2}\right) \\
&= D\left[xy + yx + x + y + \frac{1}{2}\right] - D\left(x+\frac{1}{2}\right) - D\left(y+\frac{1}{2}\right) \\
&= f(xy + yx + x + y) - f(x) - f(y)
\end{aligned}$$

for all $x, y \in R$. This means that f satisfies Eq. [4.61].

□

Theorem 4.4.6. *The mapping $f : \mathcal{A} \to \mathcal{X}$ holds in functional equation* [4.61]*, if and only if the mapping $\varphi_f : \mathcal{A} \to \mathcal{X} \oplus_1 \mathcal{A}$ defined by $\varphi_f(a) := (f(a), a)$, is a functional equation* [4.59]*. Moreover, f is continuous if and only if φ_f is continuous.*

Proof. It is similar to the proof of Theorem 4.4.5. □

4.5 Some open problems

According to Section 4.4, we raise some questions about new functional equations.

Problem 4.5.1. We know that generalized homomorphisms and generalized derivations are defined in subject categories of algebra and analysis. We can define the generalized Pompeiu function as follows: a mapping $F : R \to R$ is said to be a generalized Pompeiu function if there exists a Pompeiu function $f : R \to R$ such that $F(a + b + ab) = F(a) + F(b) + f(a)F(b)$ for all $a, b \in R$. What is the solution and stability of generalized Pompeiu's functional equation?

Similarly, we can define and ask about solution of the generalized form of functional equation [4.60].

Problem 4.5.2. Let $n \in \mathbb{N}$. In subject categories of algebra and analysis, one can find the definitions of n-homomorphisms, n-Jordan homomorphisms, and n-derivations. We can define n-Pompeiu function as follows: a mapping $f : R \to R$ is said to be an n-Pompeiu function if

$$\begin{aligned}
f(a_1 + a_2 + a_3 + \cdots + a_n + a_1a_2a_3\cdots a_n) &= f(a_1) + f(a_2) + f(a) + f(a_3) \\
&\quad + \cdots + f(a_n) + f(a_1a_2a_3\cdots a_n)
\end{aligned}$$

for all $a_1, a_2, a_3, \ldots, a_n \in R$. What is the solution and stability of n-Pompeiu's functional equation?

Similarly, we can define and ask about solution of the functional equation:

$$\begin{aligned} &f(a_1 + a_2 + a_3 + \cdots + a_n + a_1 a_2 a_3 \cdots a_n) \\ &\quad = f(a_1) + f(a_2) + f(a) + f(a_3) + \cdots + f(a_n) \\ &\qquad + f(a_1) a_2 a_3 \cdots a_n + a_1 f(a_2) a_3 a_4 \cdots a_n + a_1 a_2 f(a_3) a_4 a_5 \cdots a_n \\ &\qquad + \cdots + a_1 a_2 a_3 \cdots a_{n-1} f(a_n). \end{aligned}$$

Problem 4.5.3. Recently, John Rassias and others [73] investigated the solution and stability of a Hosszu-type functional inequality

$$f(x + y + xy) \leq f(x) + f(y) + f(xy)$$

(see also [74]). What is the solutions and stability property of the following functional inequalities for arbitrary $f, g, k, l, m, n : \mathbb{R} \to \mathbb{R}$?

$$\begin{aligned} f(x + y + z + xy + xz + yz + xyz) &\leq f(x) + f(y) + f(z) + f(x)f(y) \\ &\quad + f(x)f(z) + f(y)f(z) + f(x)f(y)f(z), \\ f(x + y + xy) &\leq g(x) + h(y) + k(x)y + xl(y), \\ f(x + y + xy + yx) &\leq g(x) + h(y) + k(x)l(y) + m(y)n(x) \end{aligned}$$

and

$$f(x + y + xy + yx) \leq g(x) + h(y) + k(x)y + xl(y) + m(y)x + yn(x).$$

We know that $\mathbb{R}$ is commutative. Then the last two inequalities can be written as

$$f(x + y + 2xy) \leq g(x) + h(y) + k(x)l(y) + +m(y)n(x)$$

and

$$f(x + y + 2xy) \leq g(x) + h(y) + k(x)y + xl(y) + m(y)x + yn(x).$$

Problem 4.5.4. What is the solution of the functional equations

$$f(x + y + xy) = f(x) + f(y) + f(x)y$$

and

$$f(x + y + xy) = f(x) + f(y) + xf(y)?$$

Problem 4.5.5. What are the solutions of the functional inequalities

$$f(x+y+xy) \leq g(x)+h(y)+k(x)y$$

and

$$f(x+y+xy) \leq g(x)+h(y)+xk(y),$$

where $f, g, h, k : \mathbb{R} \to \mathbb{R}$ are unknown functions?

Problem 4.5.6. In Section 4.4, we see that functional equations [4.58], [4.59], [4.60], and [4.61] have solutions in unital rings. If the ring has no united element, what are the solutions of these functional equations?

Stability of functional equations in inner product spaces

5

5.1 Introduction

The study on linear orthogonality preserving mappings can be considered as a part of the theory of linear preservers. In the simplest case, for X and Y being real or complex inner product spaces with the standard orthogonality relation $\perp$, a mapping $T : X \to Y$ which satisfies

$$x \perp y \Rightarrow T(x) \perp f(y), \quad x, y \in X$$

is called orthogonality preserving (o.p.). Now, if we replace the exact orthogonality $\perp$ by somehow defined approximate orthogonality $\perp^{\varepsilon}$ we obtain a larger class of approximately orthogonality preserving (a.o.p.) mappings defined by

$$x \perp y \Rightarrow f(x) \perp^{\varepsilon} f(y), \quad x, y \in X.$$

It can be proved that linear o.p. mappings are just similarities whereas linear a.o.p. ones are, in a sense, approximate similarities. Moreover, it can be shown that each linear a.o.p. mapping can be approximated by a linear o.p. one. Following the idea of the stability of functional equations, we may speak about stability of the orthogonality preserving property (see [75]).

The problem can be easily generalized from the realm of inner product spaces to normed spaces, where the norm not necessarily comes from an inner product and the orthogonality relation may be defined in various ways. Another direction of generalization is to replace the scalar-valued inner product by an inner product taking values in some C*-algebra, ie, in the realm of inner product (Hilbert) modules.

Since the o.p. mappings can be irregular, far from being continuous or linear. Thus, in considerations that follow, we restrict ourselves to linear mappings only.

Theorem 5.1.1 (Chmieliński [75]). *For a nonzero linear mapping $T : X \to Y$ the following conditions are equivalent with some $\gamma > 0$:*

1. *T preserves orthogonality;*
2. $\|T(x)\| = \gamma \|x\|, \quad x \in X;$
3. $\langle T(x)|T(y)\rangle = \gamma^2 \langle x|y\rangle, \quad x, y \in X;$ *and*
4. $|\langle T(x)|T(y)\rangle| = \gamma^2 |\langle x|y\rangle|, \quad x, y \in X.$

Theory of Approximate Functional Equations. http://dx.doi.org/10.1016/B978-0-12-803920-5.00005-5

Now, let $\mathcal{A}$ be a C*-algebra and let V, W be inner product $\mathcal{A}$-modules. We may consider $\mathcal{A}$ as a subalgebra of an algebra $\mathcal{B}(\mathcal{H})$ of all linear bounded operators on a Hilbert space $\mathcal{H}$. By $\mathcal{K}(\mathcal{H})$ we denote the subalgebra of compact operators. Ilišević and Turnšek [76] proved the validity of Theorem 5.1.3 in this setting.

Theorem 5.1.2. *For a C*-algebra $\mathcal{A}$ with $\mathcal{K}(\mathcal{H}) \subset \mathcal{A} \subset \mathcal{B}(\mathcal{H})$ and a nonzero $\mathcal{A}$-linear mapping $T : V \to W$, the following conditions are equivalent (with some $\gamma > 0$):*

Theorem 5.1.3 (Chmieliński [77]). *Let X and Y be inner product spaces and let X be finite-dimensional. Then there exists a continuous mapping $\delta : [0,1) \to \mathbb{R}_+$ such that $\lim_{\varepsilon\to 0^+} \delta(\varepsilon) = 0$ and satisfying the following property. For each mapping $f : X \to Y$ satisfying*

$$|\langle f(x)|f(y)\rangle| - \langle x|y\rangle| \le \varepsilon\|x\|\,\|y\|, \quad x, y \in X,$$

there exists a linear isometry $I : X \to Y$ such that

$$\|f(x) - I(x)\| \le \delta(\varepsilon)\|x\|, \quad x \in X.$$

5.2 Orthogonal derivations in orthogonality Banach algebras

Eshaghi and Abbaszadeh [78] investigated the stability and hyperstability of the orthogonal derivations by using the alternative of fixed point (Theorem 2.2.8). An orthogonality space $(X, \perp)$ is a real vector space X with $\dim X \ge 2$ together with a binary relation $\perp$ satisfying some axioms similar to the ones in [79].

There are several orthogonality notions on a real normed space such as Birkhoff-James, Boussouis, (semi-)inner product, Singer, Carlsson, area, unitary-Boussouis, Roberts, Pythagorean, isosceles, and Diminnie (see, eg, [80, 81]). Here, however, we present the orthogonality concept introduced by J. Rätz [82]. This is given in the following definition.

Suppose that X is a real vector space (or an algebra) with $\dim X \ge 2$ and $\perp$ is a binary relation on X with the following properties:

($\mathbf{O}_1$) totality of $\perp$ for zero: $x \perp 0$, $0 \perp x$ for all $x \in X$;

($\mathbf{O}_2$) independence: if $x, y \in X - \{0\}$, $x \perp y$, then x, y are linearly independent;

($\mathbf{O}_3$) homogeneity: if $x, y \in X$, $x \perp y$, then $\alpha x \perp \beta y$ for all $\alpha, \beta \in \mathbb{R}$; and

($\mathbf{O}_4$) the Thalesian property: if P is a two-dimensional subspace (subalgebra) of X, $x \in P$ and $\lambda \in R_+$, then there exists $u_x \in P$ such that $x \perp u_x$ and $x + u_x \perp \lambda x - u_x$.

The pair $(X, \perp)$ is called an orthogonality space (algebra). By an orthogonality normed space (normed algebra), we mean an orthogonality space (algebra) having a normed structure.

The orthogonal Cauchy functional equation $f(x+y) = f(x) + f(y)$, $x \perp y$ in which $\perp$ is an abstract orthogonality relation was first investigated in [79]. A generalized version of Cauchy's equation is the equation of Pexider type $f_1(x+y) = f_2(x) + f_3(y)$. Jun et al. [83, 90] obtained the Hyers-Ulam stability of this Pexider equation.

Let $(A, \perp)$ be an orthogonality normed algebra and B be an A-bimodule. A mapping $d : A \to B$ is an orthogonally ring derivation if d is an orthogonally additive mapping satisfying

$$d(xy) = xd(y) + d(x)y \tag{5.1}$$

for all $x, y \in A$ with $x \perp y$. Moreover, a mapping $d : A \to B$ is said to be an orthogonally Jordan ring derivation if d is an orthogonally additive mapping satisfying

$$d(xy + yx) = xd(y) + d(x)y + yd(x) + d(y)x \tag{5.2}$$

for all $x, y \in A$ with $x \perp y$. In particular, we may define orthogonally derivations associated to the Pexiderized Cauchy functional equation.

Definition 5.2.1 (Eshaghi and Abbaszadeh [78]). Let $(A, \perp)$ be an orthogonality normed algebra and B be an A-bimodule and let $f, g, h : A \to B$ be mappings satisfying the system

$$\begin{aligned} f(x+y) &= g(x) + h(y), \\ f(xy) &= xg(y) + h(x)y \end{aligned}$$

for all $x, y \in A$ with $x \perp y$, then we call it an orthogonal Pexiderized ring derivation system of equations. Moreover, if the mappings f, g, h satisfy the system

$$\begin{aligned} f(x+y) &= g(x) + h(y), \\ f(xy + yx) &= xg(y) + h(x)y + yg(x) + h(y)x \end{aligned}$$

for all $x, y \in A$ with $x \perp y$, we call it an orthogonal Pexiderized Jordan ring derivation system of equations.

In the following theorem, by applying the alternative of fixed point (Theorem 2.2.8), we will prove the Hyers-Ulam stability and hyperstability properties for the orthogonal Pexider ring derivations.

Theorem 5.2.2 (Eshaghi and Abbaszadeh [78]). *Let $\mathcal{A}$ be an orthogonality Banach algebra and $\mathcal{B}$ a Banach $\mathcal{A}$-bimodule. Suppose that $f, g, h : \mathcal{A} \to \mathcal{B}$ are mappings fulfilling the system of functional inequalities*

$$\|f(x+y) - g(x) - h(y)\| \leq \varphi(x, y), \tag{5.3}$$

$$\|f(xy) - xg(y) - h(x)y\| \leq \phi(x, y), \tag{5.4}$$

where $\varphi, \phi : X \times X \to [0, \infty)$ are mappings such that

$$\lim_{n\to\infty} \frac{\varphi(2^{nj}x, 2^{nj}y)}{2^{nj}} = 0, \tag{5.5}$$

$$\lim_{n\to\infty}\frac{\phi(2^{nj}x,y)}{2^{nj}}=\lim_{n\to\infty}\frac{\phi(x,2^{nj}y)}{2^{nj}}=0 \tag{5.6}$$

for all $x, y \in \mathcal{A}$ *with* $x \perp y$*, where* $j \in \{-1,1\}$*. If* f *is an odd mapping,* $\varphi(0,0)=\phi(0,0)=0$ *and there exists* $0<L=L(j)<1$ *such that for any fixed* $x\in\mathcal{A}$ *and some* $u_x \in \mathcal{A}$ *with* $x \perp u_x$*, the mapping*

$$\begin{aligned} x \mapsto \psi(x,u_x) = {} & \varphi\left(\frac{x+u_x}{2},\frac{x-u_x}{2}\right)+\varphi\left(0,\frac{x-u_x}{2}\right)+\varphi\left(\frac{x+u_x}{2},0\right)+\varphi\left(\frac{x}{2},\frac{u_x}{2}\right)\\ & +\varphi\left(\frac{x}{2},\frac{-u_x}{2}\right)+2\varphi\left(\frac{x}{2},0\right)+\varphi\left(0,\frac{u_x}{2}\right)+\varphi\left(0,\frac{-u_x}{2}\right) \end{aligned} \tag{5.7}$$

has the property

$$\psi(x,u_x)\le L2^j\psi\left(\frac{x}{2^j},\frac{u_x}{2^j}\right), \tag{5.8}$$

then there exists a unique orthogonally ring derivation $d:\mathcal{A}\to\mathcal{B}$ *such that*

$$\begin{aligned} \|f(x)-d(x)\| &\le \frac{L^{1+j/2}}{1-L}\psi(x,u_x),\\ \|g(x)-g(0)-d(x)\| &\le \frac{L^{1+j/2}}{1-L}\psi(x,u_x)+\varphi(x,0),\\ \|h(x)-h(0)-d(x)\| &\le \frac{L^{1+j/2}}{1-L}\psi(x,u_x)+\varphi(0,x). \end{aligned} \tag{5.9}$$

Proof. Let $E=\{e:\mathcal{A}\to\mathcal{B} \mid e(0)=0\}$. For any fixed $x\in\mathcal{A}$ and some $u_x\in\mathcal{A}$ with $x\perp u_x$, define $m:E\times E\to[0,\infty]$ by

$$m(e_1,e_2)=\inf\left\{K\in\mathbb{R}_+:\|e_1(x)-e_2(x)\|\le K\psi(x,u_x)\right\}.$$

As usual, $\inf\emptyset=\infty$. It is easy to see that (E,m) is a complete generalized metric space. Let us consider the linear mapping $T:E\to E$, $Te(x)=\frac{1}{2^j}e(2^jx)$ for all $x\in\mathcal{A}$. T is a strictly contractive mapping with the Lipschitz constant L. Indeed, for given e_1 and e_2 in E such that $m(e_1,e_2)<\infty$ and any $K>0$ satisfying $m(e_1,e_2)<K$ and any fixed $x\in\mathcal{A}$ and some $u_x\in\mathcal{A}$ with $u_{\alpha x}=\alpha u_x$ $(\alpha\in\mathbb{R})$ and $x\perp u_x$, we have

$$\begin{aligned} & \|e_1(x)-e_1(x)\|\le K\psi(x,u_x)\\ \Rightarrow & \left\|\frac{1}{2^j}e_1\left(2^jx\right)-\frac{1}{2^j}e_2\left(2^jx\right)\right\|\le\frac{1}{2^j}K\psi\left(2^jx,2^ju_x\right)\\ \Rightarrow & \left\|\frac{1}{2^j}e_1\left(2^jx\right)-\frac{1}{2^j}e_2\left(2^jx\right)\right\|\le LK\psi(x,u_x)\\ \Rightarrow & \, m\left(Te_1,Te_2\right)\le LK. \end{aligned}$$

Put $K = m(e_1, e_2) + \frac{1}{n}$ for positive integers n. Then $m(Te_1, Te_2) \leq L(m(e_1, e_2) + \frac{1}{n})$. Letting $n \to \infty$ gives

$$m(Te_1, Te_2) \leq Lm(e_1, e_2)$$

for all $e_1, e_2 \in E$.

Since $\varphi(0,0) = \phi(0,0) = 0$, putting $x, y = 0$ in Eqs. [5.3] and [5.4], we get

$$f(0) = 0, \quad g(0) + h(0) = 0. \tag{5.10}$$

For every $x, y \in \mathcal{A}$, $x, y \perp 0$. We can then put $y = 0$ and $x = 0$ in Eq. [5.3], respectively, to obtain

$$\|f(x) - g(x) - h(0)\| \leq \varphi(x, 0),$$

$$\|f(y) - g(0) - h(y)\| \leq \varphi(0, y)$$

and by Eq. [5.10], we conclude that

$$\|f(x) - \big(g(x) - g(0)\big)\| \leq \varphi(x, 0), \tag{5.11}$$

$$\|f(y) - \big(h(y) - h(0)\big)\| \leq \varphi(0, y) \tag{5.12}$$

for all $x, y \in \mathcal{A}$.

Let $x \in \mathcal{A}$ be fixed. By (O_4) there exists $u_x \in \mathcal{A}$ such that $x \perp u_x$, $x + u_x \perp x - u_x$ and $u_{\alpha x} = \alpha u_x$ for all $\alpha \in \mathbb{R}$. Hence,

$$\|f(x + u_x) - g(x) - h(u_x)\| \leq \varphi(x, u_x). \tag{5.13}$$

By (O_3), $x \perp -u_x$,

$$\|f(x - u_x) - g(x) - h(-u_x)\| \leq \varphi(x, -u_x). \tag{5.14}$$

Replacing x and y by $x + u_x$ and $x - u_x$ in Eq. [5.3], we have

$$\|f(2x) - g(x + u_x) - h(x - u_x)\| \leq \varphi(x + u_x, x - u_x). \tag{5.15}$$

Substituting $x + u_x$ for x in Eq. [5.11] and $x - u_x$ for y in Eq. [5.12], respectively, one gets the inequalities

$$\|f(x + u_x) - \big(g(x + u_x) - g(0)\big)\| \leq \varphi(x + u_x, 0), \tag{5.16}$$

$$\|f(x - u_x) - \big(h(x - u_x) - h(0)\big)\| \leq \varphi(0, x - u_x). \tag{5.17}$$

Thus, the triangle inequality and inequalities [5.15]–[5.17] yield

$$\begin{aligned}\|f(2x) - f(x+u_x) - f(x-u_x)\| &\leq \|f(2x) - g(x+u_x) - h(x-u_x)\| \\ &+ \|f(x+u_x) - \big(g(x+u_x) - g(0)\big)\| + \|f(x-u_x) - \big(h(x-u_x) - h(0)\big)\| \\ &\leq \varphi(x+u_x, x-u_x) + \varphi(x+u_x, 0) + \varphi(0, x-u_x).\end{aligned} \tag{5.18}$$

It follows from Eqs. [5.3], [5.12], [5.13], [5.14], oddness of f and triangle inequality that

$$\begin{aligned}\|2f(x) - f(x+u_x) - f(x-u_x)\| &\leq \|f(x+u_x) - g(x) - h(u_x)\| \\ &+ \|f(x-u_x) - g(x) - h(-u_x)\| + 2\|f(x) - \big(g(x) - g(0)\big)\| \\ &+ \|f(u_x) - g(0) - h(u_x)\| + \|f(-u_x) - g(0) - h(-u_x)\| \\ &\leq \varphi(x, u_x) + \varphi(x, -u_x) + 2\varphi(x, 0) + \varphi(0, u_x) + \varphi(0, -u_x).\end{aligned} \tag{5.19}$$

Now, combining Eqs. [5.18] and [5.19], we have

$$\begin{aligned}\|f(2x) - 2f(x)\| \leq\ & \varphi(x+u_x, x-u_x) + \varphi(x+u_x, 0) + \varphi(0, x-u_x) + \varphi(x, u_x) \\ & + \varphi(x, -u_x) + 2\varphi(x, 0) + \varphi(0, u_x) + \varphi(0, -u_x).\end{aligned} \tag{5.20}$$

Using Eqs. [5.7] and [5.8], we can reduce Eq. [5.20] to

$$\left\| f(x) - \frac{1}{2} f(2x) \right\| \leq \frac{1}{2} \psi(2x, 2u_x) \leq L\psi(x, u_x),$$

that is, $m(f, Tf) \leq L = L^1 < \infty$. Moreover, replacing x in Eq. [5.20] by $\frac{x}{2}$ implies the appropriate inequality for $j = -1$

$$\left\| f(x) - 2f\left(\frac{x}{2}\right) \right\| \leq \psi(x, u_x),$$

that is, $m(f, Tf) \leq 1 = L^0 < \infty$. By Theorem 2.2.8, there exists a mapping $d : \mathcal{A} \to \mathcal{B}$ which is the fixed point of T and satisfies

$$d(x) = \lim_{n\to\infty} \frac{f(2^{nj}x)}{2^{nj}},$$

since $\lim_{n\to\infty} m(T^n f, d) = 0$. The mapping d is the unique fixed point of T in the set $M = \{e \in E : m(f, e) < \infty\}$. Using Theorem 2.2.8, we get

$$m(f, d) \leq \frac{1}{1-L} m(f, Tf),$$

which yields

$$\|f(x) - d(x)\| \le \frac{L^{1+j/2}}{1-L}\psi(x, u_x).$$

Further, inequalities [5.11] and [5.12] imply that

$$\begin{aligned}\|g(x) - g(0) - d(x)\| &\le \|f(x) - \big(g(x) - g(0)\big)\| + \|f(x) - d(x)\| \\ &\le \frac{L^{1+j/2}}{1-L}\psi(x, u_x) + \varphi(x, 0), \\ \|h(x) - h(0) - d(x)\| &\le \|f(x) - \big(h(x) - h(0)\big)\| + \|f(x) - d(x)\| \\ &\le \frac{L^{1+j/2}}{1-L}\psi(x, u_x) + \varphi(0, x)\end{aligned}$$

as desired.

It follows from inequalities [5.11] and [5.12] that

$$\begin{aligned}\left\|2^{-nj}f(2^{nj}x) - 2^{-nj}\big(g(2^{nj}x) - g(0)\big)\right\| &\le 2^{-nj}\varphi(2^{nj}x, 0), \\ \left\|2^{-nj}f(2^{nj}x) - 2^{-nj}\big(h(2^{nj}x) - h(0)\big)\right\| &\le 2^{-nj}\varphi(0, 2^{nj}x)\end{aligned}$$

for all $x \in \mathcal{A}$ and $n \in \mathbb{N}$, whence

$$d(x) = \lim_{n\to\infty} \frac{g(2^{nj}x) - g(0)}{2^{nj}} = \lim_{n\to\infty} \frac{h(2^{nj}x) - h(0)}{2^{nj}}. \tag{5.21}$$

Let $x, y \in \mathcal{A}$ with $x \perp y$. (O_3) ensures $2^{nj}x \perp 2^{nj}y$ for all $n \in \mathbb{N}$ and from Eqs. [5.3], [3.148], and [5.21], we obtain

$$\begin{aligned}&\left\|2^{-nj}f\big(2^{nj}(x+y)\big) - 2^{-nj}\big(g(2^{nj}x) - g(0)\big) - 2^{-nj}\big(h(2^{nj}y) - h(0)\big)\right\| \\ &= \left\|2^{-nj}f\big(2^{nj}(x+y)\big) - 2^{-nj}g(2^{nj}x) - 2^{-nj}h(2^{nj}y)\right\| \\ &\le 2^{-nj}\varphi\left(2^{nj}x, 2^{nj}y\right).\end{aligned}$$

Therefore, from $n \to \infty$, one establishes $d(x + y) - d(x) - d(y) = 0$. Hence, d is orthogonally additive.

In addition, we claim that the mapping d satisfies the functional equation [5.1]. Define $r : \mathcal{A} \times \mathcal{A} \to \mathcal{B}$ by $r(x, y) = f(xy) - xg(y) - h(x)y$ for all $x, y \in \mathcal{A}$ with $x \perp y$. Condition [5.6] implies that

$$\lim_{n\to\infty} \frac{r(2^{nj}x, y)}{2^{nj}} = 0. \tag{5.22}$$

Utilizing the relations [5.21] and [5.22], one obtains

$$\begin{aligned} d(xy) &= \lim_{n\to\infty} \frac{f\big(2^{nj}(xy)\big)}{2^{nj}} = \lim_{n\to\infty} \frac{f\big((2^{nj}x)y\big)}{2^{nj}} \\ &= \lim_{n\to\infty} \frac{2^{nj}xg(y) + h(2^{nj}x)y + r(2^{nj}x, y)}{2^{nj}} \\ &= \lim_{n\to\infty} \left(xg(y) + \frac{h(2^{nj}x)}{2^{nj}}y + \frac{r(2^{nj}x, y)}{2^{nj}} \right) \\ &= xg(y) + d(x)y + \lim_{n\to\infty} \frac{h(0)}{2^{nj}}y. \end{aligned} \tag{5.23}$$

Set $x = 0$ in Eq. [5.23]. Since $d(0) = 0$ and $0 \perp y$ for all $y \in \mathcal{A}$, we may conclude that $\lim_{n\to\infty} \frac{h(0)}{2^{nj}}y = 0$. Hence,

$$d(xy) = xg(y) + d(x)y \tag{5.24}$$

for all $x, y \in \mathcal{A}$ with $x \perp y$.

Now, let $x, y \in \mathcal{A}$ with $x \perp y$ and $n \in \mathbb{N}$ be fixed. Using Eq. [5.24] and orthogonal additivity of d, one can easily show that

$$\begin{aligned} xg(2^{nj}y) + 2^{nj}d(x)y &= xg(2^{nj}y) + d(x)2^{nj}y = d\big(x(2^{nj}y)\big) = d\big((2^{nj}x)y\big) \\ &= 2^{nj}xg(y) + d(2^{nj}x)y = 2^{nj}xg(y) + 2^{nj}d(x)y. \end{aligned}$$

If we compare the above relation with Eq. [5.24], we get

$$x\frac{g(2^{nj}y)}{2^{nj}} = xg(y) \tag{5.25}$$

and so

$$d(xy) = x\frac{g(2^{nj}y)}{2^{nj}} + d(x)y.$$

Taking the limit as $n \to \infty$, we see that

$$d(xy) = xd(y) + \lim_{n\to\infty} x\frac{g(0)}{2^{nj}} + d(x)y. \tag{5.26}$$

Letting $y = 0$ in Eq. [5.26], we may infer that $\lim_{n\to\infty} x\frac{g(0)}{2^{nj}} = 0$. Therefore, $d(xy) = xd(y) + d(x)y$. The proof of Theorem 5.2.2 is now completed. □

In particular, given $\varphi(x, y) = \varepsilon(\|x\|^p + \|y\|^p)$ and $\phi(x, y) = \theta\|x\|^q\|y\|^s$ for $\varepsilon, \theta \geq 0$ and some real numbers p, q, s in the main theorem, one gets the following corollary (as a consequence of Rassias theorem).

Corollary 5.2.3 (Eshaghi and Abbaszadeh [78]). *Let $\mathcal{A}$ be an orthogonality Banach algebra and $\mathcal{B}$ a Banach $\mathcal{A}$-bimodule. Let $j \in \{-1, 1\}$ and $f, g, h : \mathcal{A} \to \mathcal{B}$ be mappings satisfying*

$$\|f(x+y)-g(x)-h(y)\| \leq \varepsilon(\|x\|^p+\|y\|^p),$$

$$\|f(xy)-xg(y)-h(x)y\| \leq \theta\|x\|^q\|y\|^s$$

for all $x, y \in \mathcal{A}$ with $x \perp y$, $\varepsilon, \theta \geq 0$ and real numbers p, q, s such that $p, q < 1$ for $j = 1$ and $p, q > 1$ for $j = -1$. If f is an odd mapping, then there exists a unique orthogonally ring derivation $d : \mathcal{A} \to \mathcal{B}$ such that

$$\|f(x)-d(x)\| \leq \frac{2^{j(1+j)(p-1)/2}}{1-2^{j(p-1)}}\varepsilon\left(2\|x+u_x\|^p+2\|x-u_x\|^p+4\|x\|^p+4\|u_x\|^p\right),$$

$$\|g(x)-g(0)-d(x)\|$$
$$\leq \varepsilon\left\{\frac{2^{j(1+j)(p-1)/2}}{1-2^{j(p-1)}}\left(2\|x+u_x\|^p+2\|x-u_x\|^p+4\|x\|^p+4\|u_x\|^p\right)+(\|x\|^p)\right\},$$

$$\|h(x)-h(0)-d(x)\|$$
$$\leq \varepsilon\left\{\frac{2^{j(1+j)(p-1)/2}}{1-2^{j(p-1)}}\left(2\|x+u_x\|^p+2\|x-u_x\|^p+4\|x\|^p+4\|u_x\|^p\right)+(\|x\|^p)\right\} \tag{5.27}$$

for any fixed $x \in \mathcal{A}$ and some $u_x \in \mathcal{A}$ with $x \perp u_x$.

Proof. Let $\varphi(x, y) = \varepsilon(\|x\|^p + \|y\|^p)$ and $\phi(x, y) = \theta\|x\|^q\|y\|^s$. Clearly, $\varphi(0, 0) = \phi(0, 0) = 0$. It follows from the hypotheses of the corollary that

$$\lim_{n\to\infty}\frac{\varphi(2^{nj}x, 2^{nj}y)}{2^{nj}} = \lim_{n\to\infty}\varepsilon 2^{nj(p-1)}(\|x\|^p+\|y\|^p) = 0,$$
$$\lim_{n\to\infty}\frac{\phi(2^{nj}x, y)}{2^{nj}} = \lim_{n\to\infty}\theta\, 2^{nj(q-1)}\|x\|^q\|y\|^s = 0$$

for all $x, y \in \mathcal{A}$ with $x \perp y$, that is, the conditions [3.148] and [5.6] in the Theorem 5.2.2 are sharp here. Since the inequality

$$2^{-j}\psi(2^jx, 2^ju_x) = 2^{j(p-1)}\varepsilon\left(2\|x+u_x\|^p+2\|x-u_x\|^p+4\|x\|^p+4\|u_x\|^p\right)$$
$$\leq 2^{j(p-1)}\psi(x, u_x)$$

holds for any fixed $x \in \mathcal{A}$, some $u_x \in \mathcal{A}$ with $x \perp u_x$, $\varepsilon \geq 0$ and real numbers p such that $p < 1$ for $j = 1$ and $p > 1$ for $j = -1$, we see that inequality [5.8] in Theorem 5.2.2 holds with $L = 2^{j(p-1)}$. Now, by Eq. [5.9], we conclude the assertion of this corollary. □

Next, we are going to establish the hyperstability of the orthogonal Pexider ring derivation.

Corollary 5.2.4 (Eshaghi and Abbaszadeh [78]). *Let $\mathcal{A}$ be an orthogonality Banach algebra and $\mathcal{B}$ a Banach $\mathcal{A}$-bimodule. Assume that $f, g, h : \mathcal{A} \to \mathcal{B}$ are mappings satisfying the system*

$$\|f(x+y)-g(x)-h(y)\| \leq \varphi(x,y),$$

$$\|f(xy)-xg(y)-h(x)y\| \leq \phi(x,y),$$

where $\varphi, \phi : \mathcal{A} \times \mathcal{A} \to [0,\infty)$ *are mappings such that*

$$\lim_{n\to\infty} \frac{\varphi(2^{nj}x, 2^{nj}y)}{2^{nj}} = 0,$$

$$\lim_{n\to\infty} \frac{\phi(2^{nj}x, y)}{2^{nj}} = \lim_{n\to\infty} \frac{\phi(x, 2^{nj}y)}{2^{nj}} = 0 \tag{5.28}$$

for all $x, y \in \mathcal{A}$ *with* $x \perp y$, *where* $j \in \{-1, 1\}$. *Let* $g(0) = h(0) = 0$ *and* $\mathcal{B}$ *be a Banach* $\mathcal{A}$*-bimodule without order, ie,* $\mathcal{A}x = 0$ *or* $x\mathcal{A} = 0$ *implies that* $x = 0$. *If* f *is an odd mapping,* $\varphi(0,0) = \phi(0,0) = 0$ *and there exists* $0 < L = L(j) < 1$ *such that for any fixed* $x \in \mathcal{A}$ *and some* $u_x \in \mathcal{A}$ *with* $x \perp u_x$, *the mapping* ψ *(Eq. [5.7] in Theorem 5.2.2) has the property*

$$\psi(x, u_x) \leq L2^j \psi\left(\frac{x}{2^j}, \frac{u_x}{2^j}\right),$$

then the mappings g, h *are orthogonally ring derivations. Moreover, if either* $\varphi(0,x) = 0$ *or* $\varphi(x,0) = 0$ *for all* $x \in \mathcal{A}$, *then* f *is orthogonally ring derivation.*

Proof. According to Theorem 5.2.2, there exists an orthogonally ring derivation $d : \mathcal{A} \to \mathcal{B}$ such that

$$d(x) = \lim_{n\to\infty} \frac{f(2^{nj}x)}{2^{nj}} = \lim_{n\to\infty} \frac{g(2^{nj}x)}{2^{nj}} = \lim_{n\to\infty} \frac{h(2^{nj}x)}{2^{nj}} \tag{5.29}$$

for all $x \in \mathcal{A}$, Since $g(0) = h(0) = 0$. By applying Eq. [5.29] in Eq. [5.25], we conclude that $x\big(d(y) - g(y)\big) = 0$ for all $x, y \in \mathcal{A}$. Therefore, $g = d$.

Let $x, y \in \mathcal{A}$ with $x \perp y$ and r be the mapping defined in Theorem 5.2.2. It follows from Eq. [5.28] that

$$\lim_{n\to\infty} \frac{r(x, 2^{nj}y)}{2^{nj}} = 0.$$

Using the above relation and Eq. [5.29], we obtain

$$d(xy) = xd(y) + h(x)y. \tag{5.30}$$

Similarly to the corresponding proof of Theorem 5.2.2, we have

$$\frac{h(2^{nj}x)}{2^{nj}}y = h(x)y.$$

By applying Eq. [5.29] in the previous relation, we conclude that $h = d$.

Now, we only need to show that f is orthogonally ring derivation. Applying the last hypothesis of this corollary to either relation [5.11] or relation [5.12], we indeed get the desired result. □

Theorem 5.2.5 (Eshaghi and Abbaszadeh [78]). *Let $\mathcal{A}$ be an orthogonality Banach algebra and $\mathcal{B}$ a Banach $\mathcal{A}$-bimodule. Suppose that $f, g, h : \mathcal{A} \to \mathcal{B}$ are mappings satisfying the following system of functional inequalities*

$$\|f(x+y) - g(x) - h(y)\| \leq \varphi(x,y), \tag{5.31}$$

$$\|f(xy+yx) - xg(y) - h(x)y - yg(x) - h(y)x\| \leq \phi(x,y), \tag{5.32}$$

where $\varphi, \phi : \mathcal{A} \times \mathcal{A} \to [0, \infty)$ are mappings such that

$$\lim_{n\to\infty} \frac{\varphi(2^{nj}x, 2^{nj}y)}{2^{nj}} = 0,$$

$$\lim_{n\to\infty} \frac{\phi(2^{nj}x, y)}{2^{nj}} = \lim_{n\to\infty} \frac{\phi(x, 2^{nj}y)}{2^{nj}} = 0 \tag{5.33}$$

for all $x, y \in \mathcal{A}$ with $x \perp y$, where $j \in \{-1, 1\}$. If f is an odd mapping, $\varphi(0,0) = \phi(0,0) = 0$ and there exists $0 < L = L(j) < 1$ such that for any fixed $x \in \mathcal{A}$ and some $u_x \in \mathcal{A}$ with $x \perp u_x$, the mapping ψ (Eq. [5.7] in Theorem 5.2.2) has the property

$$\psi(x, u_x) \leq L2^j \psi(\frac{x}{2^j}, \frac{u_x}{2^j}),$$

then there exists a unique orthogonally Jordan ring derivation $d : \mathcal{A} \to \mathcal{B}$ such that

$$\|f(x) - d(x)\| \leq \frac{L^{1+j/2}}{1-L}\psi(x, u_x),$$

$$\|g(x) - g(0) - d(x)\| \leq \frac{L^{1+j/2}}{1-L}\psi(x, u_x) + \varphi(x, 0),$$

$$\|h(x) - h(0) - d(x)\| \leq \frac{L^{1+j/2}}{1-L}\psi(x, u_x) + \varphi(0, x).$$

Proof. Letting $x, y = 0$ in Eqs. [5.31] and [5.32], we get

$$f(0) = 0, \quad g(0) + h(0) = 0.$$

Applying the similar argument to the corresponding part of Theorem 5.2.2, we conclude that there exists a unique orthogonally additive mapping $d : \mathcal{A} \to \mathcal{B}$, which is the fixed point of T and satisfies

$$\|f(x) - d(x)\| \le \frac{L^{1+j/2}}{1-L}\psi(x, u_x).$$

Moreover,

$$d(x) = \lim_{n\to\infty} \frac{f(2^{nj}x)}{2^{nj}} = \lim_{n\to\infty} \frac{g(2^{nj}x) - g(0)}{2^{nj}} = \lim_{n\to\infty} \frac{h(2^{nj}x) - h(0)}{2^{nj}}. \tag{5.34}$$

Now, we are going to show that the mapping d satisfies functional equation [5.2]. Define $r : \mathcal{A} \times \mathcal{A} \to \mathcal{B}$ by $r(x, y) = f(xy + yx) - xg(y) - h(x)y - yg(x) - h(y)x$ for all $x, y \in \mathcal{A}$ with $x \perp y$. It follows from Eq. [5.33] that

$$\lim_{n\to\infty} \frac{r(2^{nj}x, y)}{2^{nj}} = 0. \tag{5.35}$$

Making use of Eqs. [5.34] and [5.35], we get

$$\begin{aligned}
d(xy + yx) &= \lim_{n\to\infty} \frac{f\big(2^{nj}(xy + yx)\big)}{2^{nj}} = \lim_{n\to\infty} \frac{f\big((2^{nj}x)y + y(2^{nj}x)\big)}{2^{nj}} \\
&= \lim_{n\to\infty} \frac{2^{nj}xg(y) + h(2^{nj}x)y + yg(2^{nj}x) + h(y)2^{nj}x + r(2^{nj}x, y)}{2^{nj}} \\
&= \lim_{n\to\infty} \left(xg(y) + \frac{h(2^{nj}x)}{2^{nj}}y + y\frac{g(2^{nj}x)}{2^{nj}} + h(y)x + \frac{r(2^{nj}x, y)}{2^{nj}} \right) \\
&= xg(y) + d(x)y + yd(x) + h(y)x + \lim_{n\to\infty} \left(\frac{h(0)}{2^{nj}}y + y\frac{g(0)}{2^{nj}} \right).
\end{aligned} \tag{5.36}$$

Let $x = 0$ in Eq. [5.36]. Employing the orthogonal additivity of d and the fact that $0 \perp y$ for all $y \in \mathcal{A}$, one proves that $\lim_{n\to\infty} \left(\frac{h(0)}{2^{nj}}y + y\frac{g(0)}{2^{nj}} \right) = 0$. Hence,

$$d(xy + yx) = xg(y) + d(x)y + yd(x) + h(y)x \tag{5.37}$$

for all $x, y \in \mathcal{A}$ with $x \perp y$.

Now let $x, y \in \mathcal{A}$ with $x \perp y$ and $n \in \mathbb{N}$ be fixed. By Eq. [5.37] and orthogonal additivity of d, it can be shown that

$$\begin{aligned}
&xg(2^{nj}y) + 2^{nj}d(x)y + 2^{nj}yd(x) + h(2^{nj}y)x \\
&\quad = xg(2^{nj}y) + d(x)2^{nj}y + 2^{nj}yd(x) + h(2^{nj}y)x \\
&\quad = d\big(x(2^{nj}y) + (2^{nj}y)x\big) = d\big((2^{nj}x)y + y(2^{nj}x)\big) \\
&\quad = 2^{nj}xg(y) + d(2^{nj}x)y + yd(2^{nj}x) + h(y)2^{nj}x \\
&\quad = 2^{nj}xg(y) + 2^{nj}d(x)y + 2^{nj}yd(x) + h(y)2^{nj}x
\end{aligned}$$

and then

$$x\frac{g(2^{nj}y)}{2^{nj}} + \frac{h(2^{nj}y)}{2^{nj}}x = xg(y) + h(y)x.$$

Comparing the above relation with Eq. [5.37], we get

$$d(xy+yx) = x\frac{g(2^{nj}y)}{2^{nj}} + d(x)y + yd(x) + \frac{h(2^{nj}y)}{2^{nj}}x.$$

Sending n to infinity, we obtain

$$d(xy+yx) = xd(y) + \lim_{n\to\infty} x\frac{g(0)}{2^{nj}} + d(x)y + yd(x) + d(y)x + \lim_{n\to\infty} \frac{h(0)}{2^{nj}}x. \quad (5.38)$$

Putting $y = 0$ in Eq. [5.38], one gets $\lim_{n\to\infty} x\frac{g(0)}{2^{nj}} + \lim_{n\to\infty} \frac{h(0)}{2^{nj}}x = 0$. Hence, $d(xy+yx) = xd(y) + d(x)y + yd(x) + d(y)x$. This completes the proof of the theorem. □

As a special case, if one takes $\varphi(x,y) = \varepsilon(\|x\|^p + \|y\|^p)$ and $\phi(x,y) = \theta\|x\|^q\|y\|^s$ for $\varepsilon, \theta \geq 0$ and some real numbers p, q, s in Theorem 5.2.5, then one has the following corollary (as a consequence of Rassias theorem).

Corollary 5.2.6 (Eshaghi and Abbaszadeh [78]). *Let $\mathcal{A}$ be an orthogonality Banach algebra and $\mathcal{B}$ a Banach $\mathcal{A}$-bimodule. Let $f, g, h : \mathcal{A} \to \mathcal{B}$ be mappings satisfying*

$$\|f(x+y) - g(x) - h(y)\| \leq \varepsilon(\|x\|^p + \|y\|^p),$$

$$\|f(xy+yx) - xg(y) - h(x)y - yg(x) - h(y)x\| \leq \theta\|x\|^q\|y\|^s$$

for all $x, y \in \mathcal{A}$ with $x \perp y$, $\varepsilon, \theta \geq 0$ and real numbers p, q, s such that $p, q < 1$ for $j = 1$ and $p, q > 1$ for $j = -1$. If f is an odd mapping, then there exists a unique conditional Jordan ring derivation $d : \mathcal{A} \to \mathcal{B}$ such that Eq. [5.27] in Corollary 5.2.3 is sharp here for any fixed $x \in \mathcal{A}$ and some $u_x \in \mathcal{A}$ with $x \perp u_x$, where $j \in \{-1, 1\}$.

Proof. The proof of this corollary is omitted as it is similar to the proof of Corollary 5.2.3. □

We now present the hyperstability result concerning the orthogonal Pexider Jordan ring derivation. The proof is similar to that of Corollary 5.2.4 and we thus omit it.

Corollary 5.2.7 (Eshaghi and Abbaszadeh [78]). *Let $\mathcal{A}$ be an orthogonality Banach algebra and $\mathcal{B}$ a Banach $\mathcal{A}$-bimodule. Assume that $f, g, h : \mathcal{A} \to \mathcal{B}$ are mappings satisfying the system*

$$\|f(x+y) - g(x) - h(y)\| \leq \varphi(x,y),$$

$$\|f(xy+yx) - xg(y) - h(x)y - yg(x) - h(y)x\| \leq \phi(x,y),$$

where $\varphi, \phi : \mathcal{A} \times \mathcal{A} \to [0, \infty)$ *are mappings such that*

$$\lim_{n\to\infty} \frac{\varphi(2^{nj}x, 2^{nj}y)}{2^{nj}} = 0,$$

$$\lim_{n\to\infty} \frac{\phi(2^{nj}x, y)}{2^{nj}} = \lim_{n\to\infty} \frac{\phi(x, 2^{nj}y)}{2^{nj}} = 0$$

for all $x, y \in \mathcal{A}$ *with* $x \perp y$, *where* $j \in \{-1, 1\}$. *Let* $g(0) = h(0) = 0$ *and* $\mathcal{B}$ *be a Banach* $\mathcal{A}$*-bimodule without order, ie,* $\mathcal{A}x = 0$ *or* $x\mathcal{A} = 0$ *implies that* $x = 0$. *If* f *is an odd mapping,* $\varphi(0,0) = \phi(0,0) = 0$ *and there exists* $0 < L = L(j) < 1$ *such that for any fixed* $x \in \mathcal{A}$ *and some* $u_x \in \mathcal{A}$ *with* $x \perp u_x$, *the mapping* ψ *(Eq. [5.7] in Theorem 5.2.2) has the property*

$$\psi(x, u_x) \le L2^j \psi\left(\frac{x}{2^j}, \frac{u_x}{2^j}\right),$$

then the mappings g, h *are orthogonally Jordan ring derivations. Moreover, if either* $\varphi(0, x) = 0$ *or* $\varphi(x, 0) = 0$ *for all* $x \in \mathcal{A}$, *then* f *is orthogonally Jordan ring derivation.*

5.3 Some open problems

Eshaghi et al. [84] introduced the notion of the orthogonal sets and gave a real generalization of Banach' fixed point theorem.

Definition 5.3.1 (Eshaghi et al. [84]). Let $X \neq \emptyset$ and $\perp \subseteq X \times X$ be an binary relation. If $\perp$ satisfies the condition

$$\exists x_0; (\forall y; y \perp x_0) \quad or \quad (\forall y; x_0 \perp y),$$

it is called an orthogonal set (briefly O-set). We denote this O-set by $(X, \perp)$.

As an illustration, we consider the following examples.

Example 5.3.2 (Eshaghi et al. [84]). Let X be the set of all peoples in the world. We define $x \perp y$ if x can give blood to y. According to Table 5.1, if x_0 is a person such that

Table 5.1 **Example of blood types**

Type	You can give blood to	You can receive blood from
A+	A+ AB+	A+ A− O+ O−
O+	O+ A+ B+ AB+	O+ O−
B+	B+ AB+	B+ B− O+ O−
AB+	AB+	Everyone
A−	A+ A− AB+ AB−	A− O−
O−	Everyone	O−
B−	B+ B− AB+ AB−	B− O−
AB−	AB+ AB−	AB− B− O− A−

his (or her) blood type is O$-$, then we have $x_0 \perp y$ for all $y \in X$. This means that $(X, \perp)$ is an O-set. In this O-set, x_0 (in definition) is not unique.

Note that in the above example, x_0 may be a person with blood type AB+. In this case, we have $y \perp x_0$ for all $y \in X$.

Definition 5.3.3 (Eshaghi et al. [84]). Let $(X, \perp)$ be O-set. A sequence $\{x_n\}_{n \in \mathbb{N}}$ is called orthogonal sequence (briefly O-sequence) if

$$(\forall n; x_n \perp x_{n+1}) \quad or \quad (\forall n; \forall x_{n+1} \perp x_n).$$

Definition 5.3.4 (Eshaghi et al. [84]). Let $(X, \perp, d)$ be an orthogonal metric space ($(X, \perp)$ is an O-set and (X, d) is a metric space). Then $f : X \to X$ is orthogonally continuous ($\perp$-continuous) in $a \in X$ if for each O-sequence $\{a_n\}_{n \in \mathbb{N}}$ in X if $a_n \to a$, then $f(a_n) \to f(a)$. Also f is $\perp$-continuous on X if f is $\perp$-continuous in each $a \in X$.

Definition 5.3.5 (Eshaghi et al. [84]). Let $(X, \perp, d)$ be an orthogonal metric space and $0 < \lambda < 1$. A mapping $f : X \to X$ is said to be orthogonally contraction ($\perp$-contraction) with Lipschitz constant λ if

$$d(fx, fy) \leq \lambda d(x, y), \quad x \perp y.$$

It is easy to show that every contraction is $\perp$-contraction, but the converse is not true.

The following theorem is a real extension of Banach contraction principle.

Theorem 5.3.6 (Eshaghi et al. [84]). *Let $(X, \perp, d)$ be an O-complete metric space (not necessarily complete metric space) and $0 < \lambda < 1$. Let $f : X \to X$ be $\perp$-continuous, $\perp$-contraction (with Lipschitz constant λ) and $\perp$-preserving, then f has a unique fixed point x^* in X. Also, f is a Picard operator, that is,* $\lim f^n(x) = x^*$ *for all $x \in X$.*

We shall now present some problems concerning the new extension of Banach contraction principle.

Problem 5.3.7. Theorem 5.3.6 is an extension of Theorem 2.2.8 with new definition of orthogonality. It is now natural to expect a new extension of Theorem 2.2.8 based on Theorem 5.3.6.

Problem 5.3.8. As we noted in Chapter 2, Radu [13] showed that the theorems of Hyers, Rassias, and Gajda concerning the stability of Cauchy's functional equation in Banach spaces are direct consequences of the alternative of fixed point (Theorem 2.2.8). Are these theorems consequences of extension of Theorem 2.2.8 based on Theorem 5.3.6?

Problem 5.3.9. Problem 5.3.8 can be stated for other kinds of functional equations such as quadratic, cubic, etc., moreover, mixed types of functional equations, see Chapter 3.

Amenability of groups (semigroups) and the stability of functional equations

6

6.1 Introduction

The original theorem of Hyers (Theorem 2.1.1) holds when the mapping involved is defined on an abelian group. This yields to the following question: is this fact indeed essential? The answer is no. Székelyhidi [85] proved that the amenability of the group is enough to ensure stability. Now another question naturally arises: what are the connections between stability of homomorphisms and amenability of the group where they are defined? Forti [86] first tried to give answers to these questions.

6.2 The stability of homomorphisms and amenability

We start this section with the following definition.

Definition 6.2.1. [Forti [86]] Let G be a group (or a semigroup) and B a Banach space. We say that the couple (G, B) has the property of the stability of homomorphisms (shortly (G, B) is HS) if for every function $f : G \to B$ such that

$$\|f(xy) - f(x) - f(y)\| \leq K$$

for every $x, y \in G$ and for some K, there exist $\phi \in Hom(G, B)$ and K' depending only on K such that

$$\|f(x) - \phi(x)\| \leq K' \tag{6.1}$$

for all $x \in G$.

A glance at the proof of theorem of Hyers (Theorem 2.1.1) shows that it remains true if the Banach space E is substituted by an arbitrary abelian group or semigroup; so we can say that for all Banach spaces B and all abelian groups (or semigroups), G the couple (G, B) is HS. From Theorem 2.1.1 we easily get the following result.

Proposition 6.2.2. *[Forti [86]] Let G be an arbitrary group (or semigroup) and let B be a Banach space. Assume that $f : G \to B$ satisfies the inequality*

$$\|f(xy) - f(x) - f(y)\| \leq K$$

Theory of Approximate Functional Equations. http://dx.doi.org/10.1016/B978-0-12-803920-5.00006-7

for every $x, y \in G$. *Then the limit* $\lim_{n\to\infty} 2^{-n} f\left(x^{2^n}\right)$ *exists for all* $x \in G$ *and*

$$\|f(x) - g(x)\| \leq K \quad \text{and} \quad g\left(x^2\right) = 2g(x) \tag{6.2}$$

for all $x \in G$. *The function g is the unique satisfying conditions* [5.3].

Proof. The existence of the limit and the first of Eq. [5.3] are contained in the first part of the proof of Hyers' theorem. The second of Eq. [5.3] is an immediate consequence of the definition of g. Let now $h : G \to B$ satisfying conditions [5.3], then

$$\|f\left(x^{2^n}\right) - h\left(x^{2^n}\right)\| = \|f\left(x^{2^n}\right) - 2^n h(x)\| \leq K,$$

dividing by 2^n and letting $n \to \infty$ we have $g(x) = h(x)$. □

Proposition 6.2.2 shows that the existence of the limit $\lim_{n\to\infty} 2^{-n} f\left(x^{2^n}\right)$ depends only on the completeness of the space B. Whether the function g is additive or not depends on the group G. These simple remarks enable us to prove the following theorem.

Theorem 6.2.3. *[Forti [86]] Let the couple* (G, B) *be HS, then the smallest constant* K' *fulfilling inequality* [5.3] *is equal to K; moreover, the homomorphism* ϕ *satisfying Eq. [5.1] is unique.*

Proof. Let g be the function defined as in Proposition 6.2.2, it fulfills the conditions [5.3]. Since (G, B) is HS, there exists $\phi \in Hom(G, B)$ such that

$$\|f(x) - \phi(x)\| \leq K'$$

for all $x \in G$. So

$$\|f\left(x^{2^n}\right) - \phi\left(x^{2^n}\right)\| = \|f\left(x^{2^n}\right) - 2^n \phi(x)\| \leq K'.$$

Dividing by 2^n and taking the limit as $n \to \infty$, we have $g(x) = \phi(x)$, thus if K' is the smallest constant fulfilling Eq. [5.1], we get $K' < K$. On the other hand, if we take $f(x) = \phi(x) + c$, where $\phi \in Hom(G, B)$ and $\|c\| = K$, we get

$$\|f(xy) - f(x) - f(y)\| \|c\| = K$$

and

$$\|f(x) - \phi(x)\| = K,$$

whence the smallest constant fulfilling Eq. [5.1] must be K. The uniqueness of ϕ follows from Proposition 6.2.2. □

Corollary 6.2.4. *[Forti [86]] If G is a finite group (or semigroup), then* (G, B) *is HS for every Banach space B, and* $\phi = 0$.

Theorem 6.2.5. *[Forti [86]] Assume that the couple $(G, \mathbb{C})$ or $(G, \mathbb{R})$ is HS. Then for every complex or real Banach space B, the couple (G, B) is HS.*

Proof. Let $f : G \to B$ be such that

$$\|f(xy) - f(x) - f(y)\| \leq K$$

for all $x, y \in G$. If B' is the (topological) dual of B, for every $L \in B'$ we have

$$|L\{f(xy) - f(x) - f(y)\}| = |L\{f(xy)\} - L\{f(x)\} - L\{f(y)\}| \leq K\|L\|.$$

Thus there exists $\phi_L \in Hom(G, \mathbb{C})$ such that

$$|L\{f(x) - \phi_L(x)\}| \leq K\|L\|.$$

Define $\phi(x) = \lim_{n\to\infty} 2^{-n} f\left(x^{2^n}\right)$ (the limit exists by Proposition 6.2.2); we have $\phi_L(x) = \lim_{n\to\infty} 2^{-n} L\left(f\left(x^{2^n}\right)\right) = L(\phi(x))$, by the continuity of L. So

$$L\big(\phi(xy)\big) = \phi_L(x) + \phi_L(y) = L(\phi(x)) + L\big(\phi(y)\big)$$

for every $L \in B'$. Hence, $\phi(xy) = \phi(x) + \phi(y)$. By Proposition 6.2.2, we have that (G, B) is HS. □

From now on in this section, we assume that the couple (G, B) (G group or semigroup) be HS, and if f is a function from G into B, by $\mathcal{C}f(x, y)$ we denote the Cauchy difference $\mathcal{C}f(x, y) = f(xy) - f(x) - f(y)$. If $\mathcal{C}f(x, y)$ is bounded, by ϕ_f we intend the homomorphism approximating f as indicated in Definition 6.2.1.

We consider the relations between the range of $\mathcal{C}f$ (when bounded) and the range of $f - \phi_f$. The results achieved will be used to solve some alternative functional equations.

Definition 6.2.6. [Forti [86]] Let M be an arbitrary subset of a real or complex vector space. By $C(M)$ we denote the convex hull of M, that is, the set of all elements of the form $\sum_{i=1}^{n} \alpha_i x_i$, $\alpha_i \geq 0$, $\sum_{i=1}^{n} \alpha_i = 1$, $x_i \in M$.

The following useful theorem is a straightforward generalization (with its proof) of Theorem 2 in [87].

Theorem 6.2.7 (Forti [86]). *Assume that $\mathcal{C}f(x, y) \in M$, where M is a bounded subset of a Banach space B. Then $h(x) \in \overline{C(-M)}$, where $h = f - \phi_f$. (By $\bar{A}$ we intend the closure of A).*

Proof. Let $x \in G$ and $u = h(x)$. Then for every positive integer s we have

$$h\left(x^s\right) = su + \sum_{i=1}^{s-1} m_i, \quad m_i \in M. \tag{6.3}$$

This can be proved by induction over s. Dividing Eq. [6.3] by s and taking the limit as $s \to \infty$, we have

$$u = - \lim_{s\to\infty} s^{-1} \sum_{i=1}^{s-1} m_i$$

(the limit exists since h is bounded), hence, $u \in \overline{C(M)}$. □

As a consequence of Theorem 6.2.7, the range of $f - \phi_f$ is contained in the closed subspace of B spanned by M. A result of this kind, under different assumptions, has been obtained by Baron [88].

Theorem 6.2.8 (Baron [88]). *Let G be an abelian group, Y a vector space over $\mathbb{Q}$ and Z an arbitrary subspace of Y. A function $f : G \to Y$ satisfies the condition*

$$f(x_1 + x_2) - f(x_1) - f(x_2) \in Z$$

for all $x_1, x_2 \in G$, if and only if there exists an additive function $g : G \to Y$ such that $f(x) - g(x) \in Z$ for every $x \in G$.

It has to be noticed that the additive function g in Baron's theorem is not uniquely determined. The following theorem gives a stronger result than Theorem 6.2.7 and it will be used in order to solve some functional equations.

Theorem 6.2.9 (Forti [86]). *Assume that $\mathcal{L}f(x,y) \in M$ $(x, y \in G)$ where M is a bounded subset of B and let $h = f - \phi_f$. If ε is the identity of G, let $-h(\varepsilon) = m_0 \in M$. Then $h(x) \in \{-(m_0 + M) + \overline{C(M)}\} \bigcap \overline{C(-M)}$.*

Proof. Since $h(\varepsilon) - h(x) - h\left(x^{-l}\right) \in M$ and $-h(\varepsilon) = m_0 \in M$, we have, for some $m \in M$, $h(x) = -h\left(x^{-l}\right) - m_0 - m$, hence $h(x) \in -h\left(x^{-l}\right) - (m_0 + M)$. By Theorem 6.2.7, $-h\left(x^{-l}\right) \in \overline{C(M)}$ and $h(x) \in \overline{C(-M)}$; thus we get the desired result. □

Now, we intend to analyze for which groups G the couple (G, B) is HS. We have already remarked that, as a consequence of Hyers' theorem and of Corollary 6.2.4, the couple (G, B) is HS for all abelian and all finite groups (or semigroups), whatever the Banach space B is.

Definition 6.2.10 (Forti [86]). Let G be a group or a semigroup and $B(G)$ be the space of all bounded complex-valued functions on G, equipped with the supremum norm $\|f\|_\infty$. A linear functional m on $B(G)$ is a left invariant mean (LIM) if

(α) $m(\bar{f}) = \overline{m(f)}, f \in B(G)$;

(β) $\inf\{f(x)\} \leq m(f) \leq \sup\{f(x)\}$ for all real-valued $f \in B(G)$; and

(γ) $m({}_xf) = m(f)$ for all $x \in G$ and $f \in B(G)$, where ${}_xf(t) = f(xt)$.

Likewise, we say m is a right invariant mean if $m(f_x) = m(f)$ for all $x \in G$, where $f_x(t) = f(tx)$, and we define two-sided invariance in the usual way. Condition (β) is equivalent to $m(f) \geq 0$ if $f \geq 0$, and $m(1) = 1$, hence $\|m\| = 1$ for every mean.

The following two propositions hold (see [89]):

Proposition 6.2.11. *If G is a semigroup with a LIM and a right invariant mean on $B(G)$, then there exists a two-sided invariant mean on $B(G)$.*

Proposition 6.2.12. *If G is a group, there is a LIM on $B(G)$ if and only if there is a right invariant mean on $B(G)$. Hence, by Proposition 6.2.11, there is a two-sided invariant mean on $B(G)$.*

Definition 6.2.13. A semigroup G is left (right) amenable if there is a left (right) invariant mean on $B(G)$; if G is a group, these conditions are the same and we say that G is amenable.

The following theorem due to Székelyhidi [85] shows that amenability implies the stability of homomorphisms. A direct proof is given in [86].

Theorem 6.2.14. *Let G be a left (right) amenable semigroup, then $(G, \mathbb{C})$ is HS.*

Proof. Let $f : G \to \mathbb{C}$ be such that

$$\|f(xy) - f(x) - f(y)\| \leq K$$

for all $x, y \in G$; then, for each fixed $x \in G$, the function $f(xy) - f(y)$, as a function of y, is in $B(G)$. Let m_y be a LIM on $B(G)$ (the suffix y denotes that m_y acts on functions of the variable y) and define

$$\phi(x) = m_y\{{}_xf - f\}$$

for all $x \in G$. We have

$$\begin{aligned}\phi(xz) &= m_y\{{}_{xz}f - f\} = m_y\{{}_{xz}f - {}_xf + {}_xf - f\} = m_y\{{}_x({}_zf - f)\} + m_y\{{}_xf - f\} \\ &= m_y\{{}_zf - f\} + \phi(x) = \phi(z) + \phi(x).\end{aligned}$$

So $\phi \in Hom(G, \mathbb{C})$. Then

$$\begin{aligned}|\phi(x) - f(x)| &= |m_y\{{}_xf - f\} - f(x)| = |m_y\{{}_xf - f - f(x)\}| \\ &\leq \sup_{y \in G}|f(xy) - f(x) - f(y)| \leq K\end{aligned}$$

for all $x \in G$ and thus $(G, \mathbb{C})$ is HS. □

Corollary 6.2.15. *Let G be a left (right) amenable semigroup, then for all Banach spaces B, (G, B) is HS.*

Now a question arises naturally: are there groups or semigroups G such that (G, B) is not HS for some B? In view of Theorem 6.2.14, we must look among non-amenable groups or semigroups. The following theorem (presented by G.L. Forti to the 22nd International Symposium on Functional Equations) gives an answer.

Theorem 6.2.16. *Let $F(a, b)$ $\big(S(a, b)\big)$ be the free group (semigroup) generated by the elements a and b. The couple $(F(a, b), \mathbb{R})$ $\big((S(a, b), \mathbb{R})\big)$ is not HS.*

Proof. We shall construct a function $f : F(a, b) \to \mathbb{R}$ such that $|\mathcal{L}f(x, y)| \leq 1$ and for every $\phi \in Hom(F(a, b), \mathbb{R})$ the difference $f - \phi$ is unbounded. To do this, if $x \in F(a, b)$, we assume that the "word" x be reduced, that is it does not contain pairs of the forms $aa^{-1}, a^{-1}a, bb^{-1}, b^{-1}b$ and it is written without exponents different from 1 and -1. We now define a function $f : F(a, b) \to \mathbb{R}$ in the following way: if $r(x)$ is the number of pairs of the form ab contained in x and $s(x)$ is the number of pairs of the form $b^{-1}a^{-1}$ contained in x, then $f(x) = r(x) - s(x)$.

The function f is unbounded and for each $x, y \in F(a, b)$ we have $f(xy)-f(x)-f(y) \in \{-1, 0, 1\}$, so $|\mathcal{L}f(x, y)| \leq 1$.

Assume now that $\phi \in Hom(F(a, b), \mathbb{R})$ exists so that $f - \phi$ is bounded. ϕ is completely determined by the values $\phi(a)$ and $\phi(b)$ and f is identically zero on the subgroups generated by a and b, respectively. Hence the boundedness of $f - \phi$ on these two subgroups implies $\phi = 0$; thus $f - \phi = f$, a contradiction since f is unbounded.

If instead of the free group $F(a, b)$ we consider the free semigroup $S(a, b)$, we get the analogous result by defining $f(x) = r(x)$. □

Theorems 6.2.14 and 6.2.16 suggest studying the connections between the stability of homomorphisms and the amenability of groups or semigroups. In fact, G.L. Forti obtained a theorem giving a necessary and sufficient condition for the amenability in terms of a kind of multi-stability.

We denote with $B^r(G)$ the space of all bounded real-valued functions on G and with $B^r\mathcal{L}(G)$ the space of all real-valued functions f on G for which $\mathcal{L}f$ is bounded on $G \times G$.

Theorem 6.2.17 (Forti [86]). *Let G be a group. G is amenable if and only if for every n-tuple $f_1, f_2, \ldots, f_n \in B^r\mathcal{L}(G)$, there exist $\phi_1, \phi_2, \ldots, \phi_n \in Hom(G, \mathbb{R})$, such that $f_i - \phi_i \in B^r(G)$ and for all n-tuples $x_1, x_2, \ldots, x_n \in G$, the inequality*

$$H(x_1, x_2, \ldots, x_n) \leq \sum_{i=1}^{n} \{\phi_i(x_1) - f_i(x_i)\} \leq K(x_1, x_2, \ldots, x_n) \tag{6.4}$$

holds, where

$$H(x_1, x_2, \ldots, x_n) = \inf_{y \in G} \sum_{i=1}^{n} \mathcal{L}f_i(x_1, y),$$

$$K(x_1, x_2, \ldots, x_n) = \sup_{y \in G} \sum_{i=1}^{n} \mathcal{L}f_i(x_i, y).$$

Proof. Assume that G is amenable and let m_y be a LIM on $B(G)$. If $f_i \in B^r\mathcal{L}(G)$ we put $\phi_i(x) = m_y\{_x f_i - f_i\}$ for all $x \in G$. By Theorem 6.2.14, we have $\phi_i \in Hom(G, \mathbb{R})$ and $f_i - \phi_i \in B^r(G)$. Fix now $x_1, x_2, \ldots, x_n \in G$, by (β) of Definition 6.2.10 we have

$$\begin{aligned} \inf_{y \in G} \left\{ \sum_{i=1}^{n} (f_i(x_i y) - f_i(y)) \right\} &\leq m_y \left\{ \sum_{i=1}^{n} ({}_x f_i - f_i) \right\} \\ &\leq \sup_{y \in G} \left\{ \sum_{i=1}^{n} (f_i(x_i y) - f_i(y)) \right\}, \end{aligned}$$

hence, from the definition of ϕ_i and of f_i we get

$$\begin{aligned} \inf_{y \in G} \left\{ \sum_{i=1}^{n} \mathcal{L}f_i(x_i, y) \right\} + \sum_{i=1}^{n} f_i(x_i) &\leq \sum_{i=1}^{n} \phi_i(x_i) \\ &\leq \sup_{y \in G} \left\{ \sum_{i=1}^{n} \mathcal{L}f_i(x_i, y) \right\} + \sum_{i=1}^{n} f_i(x_i). \end{aligned}$$

Thus we get inequality [5.11].

Conversely, let $f_1, f_2, \ldots, f_n \in B^r(G)$, then $f_1, f_2, \ldots, f_n \in B^r\mathcal{L}(G)$ and the corresponding homomorphisms (existing by hypothesis) ϕ_i are equal to zero. We now show that inequality [5.11] implies the condition of Dixmier (see [89]). We must show that

$$\sup_{y\in G}\left\{\sum_{i=1}^{n}(f_i(x_i y) - f_i(y))\right\} \geq 0. \tag{6.5}$$

Assuming that the supremum in Eq. [6.5] be equal to $-\varepsilon < 0$, then

$$\sum_{i=1}^{n} \mathcal{L}f_i(x_i, y) < -\varepsilon - \sum_{i=1}^{n} f_i(x_i),$$

hence, by Eq. [5.11] (the ϕ_is are zero) we get

$$-\sum_{i=1}^{n} f_i(x_i) < -\varepsilon - \sum_{i=1}^{n} f_i(x_i),$$

a contradiction. □

6.3 Some open problems

Problem 6.3.1. It is well known that a group containing $F(a,b)$ as a subgroup is not amenable. Hence a question arises: can Theorem 6.2.16 be extended to groups containing $F(a,b)$? This can be formulated in the following way: let $G \supset F(a,b)$ and let $f : F(a,b) \to \mathbb{R}$ be defined as in Theorem 6.2.16. Is it possible to extend f to $\tilde{f} : G \to \mathbb{R}$ such that $|\mathcal{L}\tilde{f}(x,y)| \leq K$ for all $x, y \in G$?

Problem 6.3.2. Let S be a semigroup and B a Banach space. We say that the couple (S, B) has the property of the stability of Jensen functional equation (shortly (S, B) is JS) if for every function $f : S \to B$ such that

$$\|2f(xy) - f(x^2) - f(y^2)\| \leq \delta$$

for all $x, y \in S$ and for some finite $\delta \geq 0$, there exists a Jensen function $J : S \to B$ such that

$$\|f(x) - J(x)\| \leq \acute{\delta}$$

for all $x \in S$, where finite constant $\acute{\delta} \geq 0$ depends only on δ.

Let S be an Abelian semigroup and B be a Banach space. If $f : S \to B$ satisfies

$$\|2f(xy) - f(x^2) - f(y^2)\| \leq \delta$$

for some finite $\delta \geq 0$ and for all $x, y \in S$, is there a Jensen function $J : S \to B$ such that

$$\|f(x) - J(x)\| \leq \acute{\delta}$$

for all $x \in S$, where the finite constant $\acute{\delta} \geq 0$ depends only on δ?

Problem 6.3.3. Assume that S is an left (right) amenable 2-divisible semigroup and assume that $f : S \to \mathbb{C}$ is a mapping satisfying

$$|2f(xy) - f(x^2) - f(y^2)| \leq \varepsilon$$

for $\varepsilon \geq 0$ and for all $x, y \in S$, where $\mathbb{C}$ is the set of all complex numbers. Is there a Jensen function $\varphi : S \to \mathbb{C}$ such that

$$|\varphi(a) - f(a)| \leq 2\epsilon$$

for all $a \in S$?

Problem 6.3.4. The equation

$$\phi(xyz) + \phi(x) + \phi(y) + \phi(z) = \phi(xy) + \phi(yz) + \phi(xz)$$

is called the Deeba functional equation. Let S be a semigroup and X a Banach space.

Assume that S is a left amenable semigroup and assume that $f : S \to \mathbb{C}$ is a mapping satisfying

$$|f(xyz) + f(x) + f(y) + f(z) - f(xy) - f(xz) - f(yz)| \leq \delta$$

for $\delta \geq 0$ and for all $x, y \in S$. Is there a Deeba function $\varphi : S \to \mathbb{C}$ such that

$$|\varphi(x) - f(x)| \leq \delta$$

for all $x \in S$?

Problem 6.3.5. Let S be a semigroup. A mapping $\varphi : S \to \mathbb{C}$ is called multiplicative if $\varphi(xy) = \varphi(x)\varphi(y)$. We remark that φ is a nonzero function because if there exist $t \in S$ such that $\varphi(t) = 0$, then we have $\varphi(te) = \varphi(t)\varphi(e) = 0$ and

$$\varphi(z) = \varphi(ze) = \varphi(zt^{-1}t) = \varphi(zt^{-1})\varphi(t) = 0$$

for all $z \in S$. This implies that $\varphi = 0$.

Assume that S is an extremely left amenable semigroup and assume that $f : S \to \mathbb{C}$ is a mapping satisfying

$$|f(xy) - f(x)f(y)| \leq \varepsilon \min\{|f(x)|, |f(y)|\}$$

for $\varepsilon \geq 0$ and for all $x, y \in S$. Is there a homomorphism $\varphi : S \to \mathbb{C}$ such that

$$|\varphi(x) - f(x)| \leq \varepsilon$$

for all $x \in S$?

Problem 6.3.6. Let S be an extremely left amenable semigroup and A be a C^*-algebra. Let $f : S \to A$ be a mapping satisfying

$$\|f(xy) - f(x)f(y)\| \leq \varepsilon \min\{\|f(x)\|, \|f(y)\|\}$$

for $\varepsilon \geq 0$ and for all $x, y \in S$, where $f(x)$ is positive for all $x \in S$. Is there a multiplicative function $\phi : S \to A$ such that

$$\|\phi(x) - f(x)\| \leq \epsilon$$

for all $x \in S$?

Bibliography

[1] T. Aoki, On the stability of the linear transformation in Banach spaces, J. Math. Soc. Jpn. 2 (1950) 64–66.

[2] J. Baker, The stability of the cosine equation, Proc. Am. Math. Soc. 80 (1980) 411–416.

[3] J. Baker, J. Lawrence, F. Zorzitto, The stability of the equation $f(x+y)=f(x)f(y)$, Proc. Am. Math. Soc. 74 (1979) 242–246.

[4] R. Ger, Superstability is not natural, Rocznik Nauk.-Dydakt. Prace Mat. 159 (1993) 109–123.

[5] R. Ger, P. Šemrl, The stability of the exponential equation, Proc. Am. Math. Soc. 124 (1996) 779–787.

[6] D. Hyers, On the stability of the linear functional equation, Proc. Natl. Acad. Sci. U. S. A. 27 (1941) 222–224.

[7] J. Baker, The stability of certain functional equations, Proc. Am. Math. Soc. 112 (1991) 729–732.

[8] T. Rassias, On the stability of the linear mapping in Banach spaces, Proc. Am. Math. Soc. 72 (1978) 297–300.

[9] J. Rassias, On approximation of approximately linear mappings by linear mappings, J. Funct. Anal. 46 (1982) 126–130.

[10] Z. Gajda, On stability of additive mappings, Int. J. Math. Math. Sci. 14 (1991) 431–434.

[11] T. Rassias, P. Šemrl, On the behavior of mappings which do not satisfy Hyers-Ulam stability, Proc. Am. Math. Soc. 114 (1992) 989–993.

[12] G. Isac, T. Rassias, Stability of ψ-additive mappings: applications to nonlinear analysis, Int. J. Math. Math. Sci. 19 (1996) 219–228.

[13] V. Radu, The fixed point alternative and the stability of functional equations, Fixed Point Theory 4 (2003) 91–96.

[14] B. Margolis, J. Diaz, A fixed point theorem of the alternative for contractions on a generalized complete metric space, Bull. Am. Math. Soc. 74 (1968) 305–309.

[15] J. Schwaiger, 12. Remark, Report of Meeting, Aequationes Math. 35 (1988) 120–121.

[16] G. Forti, J. Schwaiger, Stability of homomorphisms and completeness, C. R. Math. Rep. Acad. Sci. Canada 11 (1989) 215–220.

[17] Z. Moszner, Stability of the equation of homomorphism and completeness of the underlying space, Opusc. Math. 28 (2008) 83–92.

[18] A. Najati, On the completeness of normed spaces, Appl. Math. Lett. 23 (2010) 880–882.

[19] A. Fošner, R. Ger, A. Gilanyi, M. Moslehian, On linear functional equations and completeness of normed spaces, Banach J. Math. Anal. 7 (2013) 196–200.

[20] J. Aczél, J. Dhombres, Functional Equations in Several Variables, Cambridge University Press, Cambridge 1989.

[21] T. Rassias, On a modified Hyers-Ulam sequence, J. Math. Anal. Appl. 158 (1991) 106–113.

[22] K. Jun, H. Kim, The generalized Hyers-Ulam-Rassias stability of a cubic functional equation, J. Math. Anal. Appl. 274 (2002) 867–878.

[23] I.-S. Chang, Y.-S. Jung, Stability of a functional equation deriving from cubic and quadratic functions, J. Math. Anal. Appl. 283 (2003) 491–500.

[24] K. Jun, H. Kim, Ulam stability problem for a mixed type of cubic and additive functional equation, Bull. Belg. Math. Soc. Simon Stevin 13 (2006) 271–285.
[25] A. Najati, M. Moghimi, Stability of a functional equation deriving from quadratic and additive functions in quasi-Banach spaces, J. Math. Anal. Appl. 337 (2008) 399–415.
[26] W.-G. Park, J.-H. Bae, On a bi-quadratic functional equation and its stability, Nonlinear Anal. 62 (2005) 643–654.
[27] M. Eshaghi Gordji, S. Abbaszadeh, C. Park, On the stability of a generalized quadratic and quartic type functional equation in quasi-Banach spaces, J. Inequal. Appl. 2009 (2009), Article ID 153084, 26 pages.
[28] M. Eshaghi Gordji, H. Khodaei, Solution and stability of generalized mixed type cubic, quadratic and additive functional equation in quasi-Banach spaces, Nonlinear Anal. 71 (2009) 5629–5643.
[29] S. Rolewicz, Metric Linear Spaces, PWN-Polish Scientific Publishers, Warszawa, Reidel, Dordrecht, 1984.
[30] Y. Benyamini, J. Lindenstrauss, Geometric Nonlinear Functional Analysis, American Mathematical Society, Providence, RI, 1998.
[31] J. Tabor, stability of the Cauchy functional equation in quasi-Banach spaces, Ann. Polon. Math. 83 (2004) 243–255.
[32] M. Eshaghi Gordji, A. Ebadian, S. Zolfaghari, Stability of a functional equation deriving from cubic and quartic functions, Abstr. Appl. Anal. 2008 (2008), Article ID 801904, 17 pages.
[33] H. Kim, On the stability problem for a mixed type of quartic and quadratic functional equation, J. Math. Anal. Appl. 324 (2006) 358–372.
[34] A. Najati, G. Zamani Eskandani, Stability of a mixed additive and cubic functional equation in quasi-Banach spaces, J. Math. Anal. Appl. 342 (2008) 1318–1331.
[35] K. Hensel, Über eine neue Begründung der Theorie der algebraischen Zahlen, Jahresber. Deutsch. Math. Verein. 6 (1897) 83–88.
[36] L. Arriola, W. Beyer, Stability of the Cauchy functional equation over p-adic fields, Real Anal. Exch. 31 (2005) 125–132.
[37] M. Eshaghi Gordji, R. Khodabakhsh, S.-M. Jung, H. Khodaei, AQCQ-functional equation in non-Archimedean normed spaces, Abstr. Appl. Anal. 2010 (2010), Article ID 741942.
[38] D. Bourgin, Classes of transformations and bordering transformations, Bull. Am. Math. Soc. 57 (1951) 223–237.
[39] D. Bourgin, Approximately isometric and multiplicative transformations on continuous function rings, Duke Math. J. 16 (1949) 385–397.
[40] R. Badora, On approximate ring homomorphisms, J. Math. Anal. Appl. 276 (2002) 589–597.
[41] M. Eshaghi Gordji, A. Najati, Approximately J-homomorphisms: a fixed point approach, J. Geom. Phys. 60 (2010) 809–814.
[42] R. Badora, On approximate derivations, Math. Inequal. Appl. 9 (2006) 167–173.
[43] H. Dales, Banach Algebras and Automatic Continuity, London Mathematical Society Monographs, 24, Clarendon Press, Oxford 2000.
[44] M. Eshaghi Gordji, M. Filali, Arens regularity of module actions, Stud. Math. 181 (2007) 237–254.
[45] B. Johnson, Approximately multiplicative maps between Banach algebras, J. Lond. Math. Soc. 2 (1988) 294–316.
[46] C. Park, Lie $*$-homomorphisms between Lie C^*-algebras and Lie $*$-derivations on Lie C^*-algebras, J. Math. Anal. Appl. 293 (2004) 419–434.

[47] C. Park, D.-H. Boo, J. An, Homomorphisms between C^*-algebras and linear derivations on C^*-algebras, J. Math. Anal. Appl. 337 (2008) 1415–1424.
[48] M. Gordji, H. Khodaei, A fixed point technique for investigating the stability of (α, β, γ)-derivations on Lie C^*-algebras, Nonlinear Anal. 76 (2013) 52–57.
[49] P. Novotný, J. Hrivnák, On (α, β, γ)-derivations of Lie algebras and corresponding invariant functions, J. Geom. Phys. 58 (2008) 208–217.
[50] A. Najati, A. Ranjbari, Stability of homomorphisms for a 3D Cauchy-Jensen type functional equation on C^*-ternary algebras, J. Math. Anal. Appl. 341 (2008) 62–79.
[51] M. Eshaghi Gordji, Nearly involutions on Banach algebras: a fixed point approach, J. Math. Phys. 14 (2013) 117–124.
[52] A. Cayley, On the 34 concomitants of the ternary cubic, Am. J. Math. 4 (1881) 1–15.
[53] M. Kapranov, I. Gelfand, A. Zelevinskii, Discriminants Resultants and Multidimensional Determinants, Birkhäuser, Berlin, 1994.
[54] R. Kerner, The cubic chessboard: geometry and physics, Classical Quantum Gravity 14 (1997) A203.
[55] R. Kerner, Ternary Algebraic Structures and Their Applications in Physics, 2000, arXiv preprint math-ph/0011023.
[56] V. Abramov, R. Kerner, B. Le Roy, Hypersymmetry a Z_3 graded generalization of supersymmetry, J. Math. Phys. 38 (1997) 1650–1669.
[57] Y. Daletskii, L. Takhtajan, Leibniz and Lie algebra structures for Nambu algebra, Lett. Math. Phys. 39 (1997) 127–141.
[58] L. Takhtajan, On foundation of the generalized Nambu mechanics, Commun. Math. Phys. 160 (1994) 295–315.
[59] L. Vainerman, R. Kerner, On special classes of n-algebras, J. Math. Phys. 37 (1996) 2553–2565.
[60] S. Duplij, Ternary Hopf algebras, Symmetry Nonlinear Math. Phys. 2 (2002) 439–448.
[61] N. Bazunova, A. Borowiec, R. Kerner, Universal differential calculus on ternary algebras, Lett. Math. Phys. 67 (2004) 195–206.
[62] G. Sewell, Quantum Mechanics and its Emergent Macrophysics, Princeton University Press, Princeton, NJ, 2002.
[63] R. Haag, D. Kastler, An algebraic approach to quantum field theory, J. Math. Phys. 5 (1964) 848–861.
[64] F. Bagarello, G. Morchio, Dynamics of mean-field spin models from basic results in abstract differential equations, J. Stat. Phys. 66 (1992) 849–866.
[65] L. Cădariu, V. Radu, On the stability of the Cauchy functional equation: a fixed point approach, Grazer Math. Ber. 346 (2004) 43–52.
[66] L. Harris, Bounded Symmetric Homogeneous Domains in Infinite-Dimensional Spaces, Lecture Notes in Mathematics, Springer, Berlin 1974.
[67] M. Elin, L. Harris, S. Reich, D. Shoikhet, Evolution equations and geometric function theory in J^*-algebras, J. Nonlinear Convex Anal. 3 (2002) 81–121.
[68] L. Harris, Operator Siegel domains, Proc. R. Soc. Edinb. A 79 (1977) 137–156.
[69] M. Eshaghi Gordji, M. Ghaemi, S. Kaboli Gharetapeh, S. Shams, A. Ebadian, On the stability of J^*-derivations, J. Geom. Phys. 60 (2010) 454–459.
[70] P. Kannappan, P. Sahoo, On the generalization of the Pompeiu functional equation, Int. J. Math. Math. Sci. 21 (1998) 117–124.
[71] E. Koh, The Cauchy functional equations in distributions, Proc. Am. Math. Soc. 106 (1989) 641–646.
[72] M. Eshaghi Gordji, M. Moslehian, A trick for investigation of approximate derivations, Math. Commun. 15 (2010) 99–105.

[73] B. Bouikhalene, J. Rassias, A. Charifi, S. Kabbaj, On the approximate solution of Hosszú's functional equation, Int. J. Nonlinear Anal. Appl. 3 (2012) 40–44.
[74] W. Fechner, On a question of J.M. Rassias, Bull. Aust. Math. Soc. 89 (2014) 494–499.
[75] J. Chmieliński, Orthogonality preserving property and its Ulam stability, in: Functional Equations in Mathematical Analysis, Springer, New York, 2012.
[76] D. Ilišević, A. Turnšek, Approximately orthogonality preserving mappings on C^*-modules, J. Math. Anal. Appl. 341 (2008) 298–308.
[77] J. Chmieliński, Stability of the orthogonality preserving property in finite-dimensional inner product spaces, J. Math. Anal. Appl. 318 (2006) 433–443.
[78] M. Eshaghi, S. Abbaszadeh, On the orthogonal Pexider derivations in orthogonality Banach algebras, Fixed Point Theory (in press).
[79] S. Gudder, D. Strawther, Orthogonally additive and orthogonally increasing functions on vector spaces, Pac. J. Math. 58 (1975) 427–436.
[80] J. Alonso, C. Benítez, Carlos orthogonality in normed linear spaces: a survey. II. Relations between main orthogonalities, Extracta Math. 4 (1989) 121–131.
[81] J. Alonso, C. Benítez, Orthogonality in normed linear spaces: a survey. I. Main properties, Extracta Math. 3 (1988) 1–15.
[82] J. Rätz, On orthogonally additive mappings, Aequationes Math. 28 (1985) 35–49.
[83] Y.-H. Lee, K.-W. Jun, A generalization of the Hyers-Ulam-Rassias stability of Pexider equation, J. Math. Anal. Appl. 246 (2000) 627–638.
[84] M. Eshaghi, M. Ramezani, Y. Cho, de la Sen M., On orthogonal sets and Banach fixed point theorem, Fixed Point Theory (in press).
[85] L. Székelyhidi, Fréchet's equation and Hyers theorem on noncommutative semigroups, Ann. Polon. Math. 2 (1988) 183–189.
[86] G. Forti, The stability of homomorphisms and amenability with applications to functional equations, Abh. Math. Sem. Univ. Hamburg 57 (1987) 215–226.
[87] G. Forti, On an alternative functional equation related to the Cauchy equation, Aequationes Math. 24 (1982) 195–206.
[88] K. Baron, Functions with differences in subspaces, in: Proceedings of the 18th International Symposium on Functional Equations, University of Waterloo, Faculty of Mathematics, Waterloo, Ontario, Canada, 1980.
[89] F.P. Greenleaf, Invariant means on topological groups, Van Nostrand Mathematical Studies, vol. 16, New York, Toronto, London, Melbourne, 1969.
[90] S. Ulam, A Collection of Mathematical Problems, Interscience Publisher, New York, 1960.
[91] K. Jun, D. Shin, B. Kim, On the Hyers-Ulam-Rassias stability of the Pexider equation, J. Math. Anal. Appl. 239 (1999) 20–29.

Index

A

Amenability
  open problems, 127
  stability of, 127
Approximate Cauchy functional equations
  completeness and, 14-19
  Hyers theorem, 5-8
  Themistocles M. Rassias theorem, 9-14
Approximate homomorphisms
  Banach algebras, 69-78
  C^*-algebras, 78-87

B

Banach algebras
  approximate homomorphisms and derivations, 69-78, 108-110
  Hosszu-type functional inequality, stability, 108-110
  Pompeiu's functional equation, sufficient condition, 102-103
Banach bimodule over a Banach algebra, 73-76, 102
Banach contraction principle, 14, 125
Banach space, 22, 23, 27
Binary mixtures, functional equations, 21-29
Binary operation, 87-88

C

C^*-algebras, 78-87
Cauchy functional equations
  completeness and, 14-19
  Hyers theorem, 5-8
  Themistocles M. Rassias theorem, 9-14
Completeness, normed spaces, 14-19
Contractively subhomogeneous, 80, 81-82, 86
C^*-ternary algebras, 87-100
  approximately J^*-homomorphisms, 88-94
  J^*-derivations stability, 94-100
Cubic functional equation, 21-29

E

Expansively superhomogeneous, 80, 81-82, 83

F

Functional equations, 100-108
  amenability and stability, 127
  Banach algebras (*see* Banach algebras)
  binary mixtures, 21-29
  Cauchy (*see* Cauchy functional equations)
  inner product spaces, 111-126
  mixed foursome, 50-108
  ternary mixtures, 29-50

H

Homomorphisms
  Banach algebras, 69-78
  C^*-algebras, 78-87
  J^*-homomorphisms, 88-94
  open problems, 127
  stability of, 127
Hosszu-type functional inequality, 108-110
Hyers theorem, 5-8
Hyers-Ulam stability, 2-3

I

Inner product spaces, 111, 112
  open problems, 124-125
  orthogonality Banach algebras, orthogonal derivations, 112-124

J

J^*-derivations, 94-100
J^*-derivations stability, 94-100
Jensen-type functional equation, 88-90, 94-97
J^*-homomorphisms, 88-94

L

Lie C^*-algebras, 78-79, 80, 82-87
Lie (α, β, γ)-derivatives, 80, 83, 85, 86
Linear transformation, 1, 2, 5-7

M

Mixed foursome, functional equations, 50-108

N

"Nambu mechanics", 87
Non-Archimedean normed spaces, 50-108
Nonions, algebra of, 87
Normed spaces, Completeness, 14-19

O

Orthogonal derivations, orthogonality Banach algebras, 112-124
Orthogonality Banach algebras, orthogonal derivations, 112-124

P

Pompeiu's functional equation, 100-101, 102-103, 105-110

Q

Quadratic equation, 21-29
Quasi-norm, 32

T

Ternary algebraic operations, 87-100
Ternary mixtures, functional equations, 29-50
Themistocles M. Rassias theorem, 9-14
Theorem of Hyers, 5-8
Theorem of Themistocles M. Rassias, 9-14

U

Ulam, Stanislaw M., 1

Printed in the United States
By Bookmasters